Android Studio Essentials

A fast-paced guide to get you up and running with Android application development using Android Studio

Belén Cruz Zapata

BIRMINGHAM - MUMBAI

Android Studio Essentials

Copyright © 2015 Packt Publishing

All rights reserved. No part of this book may be reproduced, stored in a retrieval system, or transmitted in any form or by any means, without the prior written permission of the publisher, except in the case of brief quotations embedded in critical articles or reviews.

Every effort has been made in the preparation of this book to ensure the accuracy of the information presented. However, the information contained in this book is sold without warranty, either express or implied. Neither the author, nor Packt Publishing, and its dealers and distributors will be held liable for any damages caused or alleged to be caused directly or indirectly by this book.

Packt Publishing has endeavored to provide trademark information about all of the companies and products mentioned in this book by the appropriate use of capitals. However, Packt Publishing cannot guarantee the accuracy of this information.

First published: January 2015

Production reference: 1230115

Published by Packt Publishing Ltd.
Livery Place
35 Livery Street
Birmingham B3 2PB, UK.

ISBN 978-1-78439-720-3

www.packtpub.com

Credits

Author
Belén Cruz Zapata

Reviewers
Karan Kedar Balkar
Rick Boyer
Ankit Garg
Antonio Hernández Niñirola

Commissioning Editor
Amarabha Banerjee

Acquisition Editor
Richard Brookes-Bland

Content Development Editor
Sriram Neelakantan

Technical Editors
Mrunal M. Chavan
Dennis John

Copy Editor
Vikrant Phadke

Project Coordinator
Judie Jose

Proofreaders
Simran Bhogal
Kevin McGowan

Indexer
Monica Ajmera Mehta

Graphics
Abhinash Sahu

Production Coordinator
Conidon Miranda

Cover Work
Conidon Miranda

About the Author

Belén Cruz Zapata received her engineer's degree in computer science from the University of Murcia, Spain, where she specialized in software technology and intelligent and knowledge-based technology. She earned an MSc degree in computer science and is now working on her PhD in the software engineering research group at the University of Murcia.

During the academic year of 2013-2014, Belén collaborated with Université Mohammed V-Soussi, Rabat, Morocco. Her research was focused on mobile technologies in general but especially applied to medicine.

Belén is currently working as a mobile developer for Android and iOS in the San Francisco Bay Area. She is the author of *Testing and Securing Android Studio Applications*, *Packt Publishing*.

She maintains a blog at `http://www.belencruz.com`, where you can follow her projects. You can also follow her on Twitter at `@belen_cz`.

> I would like to thank Packt Publishing for offering me the opportunity to write this book. I would particularly like to thank Richard Brookes-Bland and Sriram Neelakantan for their valuable help.
>
> I would also like to thank my mentors during the last few months, Miguel R. and P. Salinas; my friends, especially Ana, Nerea, and the Yupi group, for cheering me up; my family, especially my parents and brother for supporting me; and finally, my significant other for everything.

About the Reviewers

Karan Kedar Balkar has been working as an independent Android application developer for the last 4 years. Born and brought up in Mumbai, he holds a bachelor's degree in computer engineering. He has written over 50 programming tutorials on his personal blog (http://karanbalkar.com), covering popular technologies and frameworks.

At present, Karan is a software engineer. He has been trained on various technologies such as Java, Oracle, and .NET. Besides being passionate about technology, he loves to write poems and travel to different places. He also likes listening to music and enjoys playing the guitar.

> Firstly, I would like to thank my parents for their constant support and encouragement. I would also like to thank my friends, Srivatsan Iyer, Ajit Pillai, and Prasaanth Neelakandan, for always inspiring and motivating me.
>
> I would like to express my deepest gratitude to Packt Publishing for giving me a chance to be a part of the reviewing process.

Rick Boyer began programming when he was 11 and wrote his first paid program before graduating from high school. Against his better judgment, programming became his career, and he never looked back. With over 20 years of professional software development experience in Windows, the Web, and several mobile platforms, he started his own software consulting business called NightSky Development. He's always had a passion for mobile computing and now focuses on Android development. His hobbies include astronomy, computer games, and gardening. You can contact him at about.me\RickBoyer.

Ankit Garg is a mobile engineer with four and a half years of work experience and is based at Mountain View, California. Currently, he works with AOL as an Android engineer. He has worked on AOL Mail Mobile Web and other Android products. He is passionate about mobile technology and user product experience.

Antonio Hernández Niñirola is a computer science engineer and mobile application developer. He was born and raised in Murcia in the southeast of Spain. He has developed several websites and mobile applications that have been published in both Google Play Market and Apple Store.

As soon as Antonio got his first smartphone—a second-hand, first-generation iPhone—he started programming small applications as a form of entertainment. What started as a hobby became a passion and is now leading his career, both professionally and academically.

After getting his BSc in computer science, he got a master's degree in technology and informatics. Antonio went for further studies and is now a doctorate student in the software engineering group of the Faculty of Computer Science of the University of Murcia. His main research topic is the usability and security assessment of mobile applications.

www.PacktPub.com

Support files, eBooks, discount offers, and more

For support files and downloads related to your book, please visit www.PacktPub.com.

Did you know that Packt offers eBook versions of every book published, with PDF and ePub files available? You can upgrade to the eBook version at www.PacktPub.com and as a print book customer, you are entitled to a discount on the eBook copy. Get in touch with us at service@packtpub.com for more details.

At www.PacktPub.com, you can also read a collection of free technical articles, sign up for a range of free newsletters and receive exclusive discounts and offers on Packt books and eBooks.

https://www2.packtpub.com/books/subscription/packtlib

Do you need instant solutions to your IT questions? PacktLib is Packt's online digital book library. Here, you can search, access, and read Packt's entire library of books.

Why subscribe?

- Fully searchable across every book published by Packt
- Copy and paste, print, and bookmark content
- On demand and accessible via a web browser

Free access for Packt account holders

If you have an account with Packt at www.PacktPub.com, you can use this to access PacktLib today and view 9 entirely free books. Simply use your login credentials for immediate access.

Table of Contents

Preface	**1**
Chapter 1: Installing and Configuring Android Studio	**5**
Preparing for installation	**5**
Downloading Android Studio	**6**
Installing Android Studio	6
Running Android Studio for the first time	7
Configuring the Android SDK	**8**
Summary	**11**
Chapter 2: Starting a Project	**13**
Creating a new project	**14**
Configuring the project	14
Selecting the form factors	15
Choosing the activity type	16
Summary	**21**
Chapter 3: Navigating a Project	**23**
The project navigation panel	**24**
The project structure	**26**
The resources folder	27
Gradle	28
Project settings	**29**
Summary	**30**
Chapter 4: Using the Code Editor	**31**
Customizing the editor settings	**32**
Code completion	**34**
Code generation	**37**
Navigating code	**37**
Useful shortcuts	**40**
Summary	**40**

Table of Contents

Chapter 5: Creating User Interfaces — 41
- The graphical editor — 42
- The text-based editor — 44
- Creating a new layout — 44
- Adding components — 45
- Supporting multiple screens — 47
- Changing the UI theme — 50
- Handling events — 51
- Summary — 54

Chapter 6: Tools — 55
- The SDK Manager — 56
- The AVD Manager — 57
- The Navigation Editor — 62
- Generating a Javadoc — 65
- Version control systems — 66
- Summary — 68

Chapter 7: Google Play Services — 69
- How Google Play services work — 69
- Services available — 70
- Adding Google Play services to Android Studio — 71
- Google Maps Android API v2 — 74
- Google+ Platform for Android — 76
- Google Play In-App Billing v3 — 77
- Google Cloud Messaging — 77
- Summary — 78

Chapter 8: Debugging — 79
- Running and debugging — 79
  - Console — 80
  - Debugger — 81
  - LogCat — 83
  - Memory Monitor — 85
- Android Device Monitor — 85
  - Threads — 86
    - Method profiling — 86
  - Heap — 88
  - Allocation Tracker — 89
  - Network Statistics — 90

File Explorer	90
Emulator Control	90
System Information	90
Summary	**91**
Chapter 9: Preparing for Release	**93**
Understanding an APK file	93
Steps to take before releasing your app	95
Generating a signed APK	96
Summary	**97**
Appendix: Getting Help	**99**
Getting help from Android Studio	99
Android online documentation	100
Updates	102
Summary	**103**
Index	**105**

Preface

Mobile applications have seen a huge increase in popularity in the last few years, and this interest is still growing among users. Mobile operating systems are available not only for smartphones but also for tablets, thus increasing the possible market quota for these applications.

Android has characteristics that make it pleasant to developers, such as its open source nature and a certain level of community-driven development. Android has always been contesting with iOS (the Apple mobile system) in everything, and with Xcode, iOS presented itself as a more centralized development environment. The new IDE, Android Studio, makes this centralization finally available for Android developers, and makes this tool indispensable for a good Android developer.

This book shows users how to develop and build Android applications with this new IDE. It is not only a "getting started" book but also a guide to advanced developers to build their applications faster and more productively. This book will follow a tutorial-like approach, from the basic features to the steps to build for release, including practical examples.

What this book covers

Chapter 1, *Installing and Configuring Android Studio*, describes the installation and basic configuration of Android Studio.

Chapter 2, *Starting a Project*, shows how to create a new project and the type of activities we can select.

Chapter 3, *Navigating a Project*, explores the basic structure of a project in Android Studio.

Chapter 4, *Using the Code Editor*, exposes the basic features of the code editor in order to get the best out of it.

Chapter 5, *Creating User Interfaces*, focuses on the creation of the user interfaces using both the graphical view and the text-based view.

Chapter 6, *Tools*, introduces the currently existing Google Play services and shows how to integrate them with a project in Android Studio.

Chapter 7, *Google Play Services*, exposes some additional tools such as Android SDK tools, Javadoc, and version control integration.

Chapter 8, *Debugging*, shows in detail how to debug an application in Android Studio and the provided information when debugging.

Chapter 9, *Preparing for Release*, describes how to prepare your application for its release.

Appendix, *Getting Help*, teaches you how to get help using Android Studio and provides a list of online sites to learn more about the topics seen in this book.

What you need for this book

For this book, you need a Windows, Mac, or Linux system. You will also need to have Java installed in your system.

Who this book is for

This book is not only a "getting started" book but also a guide for advanced developers who have not used Android Studio to build their Android apps before. This book is great for developers who want to learn the key features of Android Studio and for developers who want to create their first app.

It's assumed that you are familiar with the object-oriented programming paradigm and the Java programming language. It is also required to understand the main characteristics of the Android mobile system.

Conventions

In this book, you will find a number of styles of text that distinguish between different kinds of information. Here are some examples of these styles, and an explanation of their meaning.

Code words in text, database table names, folder names, filenames, file extensions, pathnames, dummy URLs, user input, and Twitter handles are shown as follows: "The default installation directory is `/Applications/Android\ Studio.app`."

A block of code is set as follows:

```
protected void onCreate(Bundle savedInstanceState) {
    super.onCreate(savedInstanceState);
    setContentView(R.layout.activity_main);
```

When we wish to draw your attention to a particular part of a code block, the relevant lines or items are set in bold:

```
protected void onCreate(Bundle savedInstanceState) {
    super.onCreate(savedInstanceState);
    setContentView(R.layout.activity_main);

    if (savedInstanceState != null) {
        System.out.println("savedInstanceState = " +
savedInstanceState);
    }
}
```

New terms and **important words** are shown in bold. Words that you see on the screen, in menus or dialog boxes for example, appear in the text like this: " Select **Blank Activity** and click on **Next**."

> Warnings or important notes appear in a box like this.

> Tips and tricks appear like this.

Reader feedback

Feedback from our readers is always welcome. Let us know what you think about this book—what you liked or disliked. Reader feedback is important for us as it helps us develop titles that you will really get the most out of.

To send us general feedback, simply e-mail feedback@packtpub.com, and mention the book's title in the subject of your message.

If there is a topic that you have expertise in and you are interested in either writing or contributing to a book, see our author guide at www.packtpub.com/authors.

Customer support

Now that you are the proud owner of a Packt book, we have a number of things to help you to get the most from your purchase.

Errata

Although we have taken every care to ensure the accuracy of our content, mistakes do happen. If you find a mistake in one of our books—maybe a mistake in the text or the code—we would be grateful if you could report this to us. By doing so, you can save other readers from frustration and help us improve subsequent versions of this book. If you find any errata, please report them by visiting http://www.packtpub.com/submit-errata, selecting your book, clicking on the **Errata Submission Form** link, and entering the details of your errata. Once your errata are verified, your submission will be accepted and the errata will be uploaded to our website or added to any list of existing errata under the Errata section of that title.

To view the previously submitted errata, go to https://www.packtpub.com/books/content/support and enter the name of the book in the search field. The required information will appear under the **Errata** section.

Piracy

Piracy of copyrighted material on the Internet is an ongoing problem across all media. At Packt, we take the protection of our copyright and licenses very seriously. If you come across any illegal copies of our works in any form on the Internet, please provide us with the location address or website name immediately so that we can pursue a remedy.

Please contact us at copyright@packtpub.com with a link to the suspected pirated material.

We appreciate your help in protecting our authors and our ability to bring you valuable content.

Questions

If you have a problem with any aspect of this book, you can contact us at questions@packtpub.com, and we will do our best to address the problem.

Installing and Configuring Android Studio

The new and official Google IDE, Android Studio, with all its varied features, is ready to be explored. Would you like to make your own Android applications and make these applications available to other users on Google Play Store? Can you do this easily? How can you achieve this?

This chapter will show you how to prepare your new Android Studio installation, and help you take your first steps in the new environment. We will begin by preparing the system for the installation and downloading the required files. Then we will see the welcome screen that prompts when running Android Studio for the first time, and accordingly, we will configure the Android **software development kit (SDK)** properly so that we have everything ready to create our first application.

These are the topics that we will cover in this chapter:

- Installing Android Studio
- Running Android Studio for the first time
- Configuring the Android SDK

Preparing for installation

A prerequisite to start working with Android Studio is to have Java installed on your system. The system must also be able to find the Java installation. This can be achieved by installing the **Java Development Kit (JDK)** on your system and then setting an environment variable named JAVA_HOME, which points to the SDK folder in your system. Check this environment variable to avoid issues during the installation of Android Studio.

Installing and Configuring Android Studio

Downloading Android Studio

The Android Studio package can be downloaded from the Android developer tools web page, at `http://developer.android.com/sdk/index.html`, by clicking on the download button, as is shown in the next screenshot. This package is an EXE file for Windows systems, a DMG file for Mac OS X systems, and a TGZ file for Linux systems.

Installing Android Studio

In Windows, launch the EXE file. The default installation directory is `\Users\<your_user_name>\AppData\Local\Android\android-studio`. The `AppData` directory is usually a hidden directory.

In Mac OS X, open the DMG file and drop Android Studio into your `Applications` folder. The default installation directory is `/Applications/Android\ Studio.app`.

In Linux systems, unzip the TGZ file and execute the `studio.sh` script located in the `android-studio/bin/` directory.

If you have any problem in the installation process or in the following steps, you can get help on this and on other known issues by checking out *Appendix, Getting Help*.

Running Android Studio for the first time

Execute Android Studio and wait until it loads completely. This may take a few minutes on the first time. The first time you execute Android Studio, you will be prompted by a welcome screen. As shown in the following screenshot, the welcome screen includes a section to open recent projects and the **Quick Start** section. The **Quick Start** section provides options to start a new project, open a project, import a project, and even perform more advanced actions such as checking out from a version control system and modifying the configuration options.

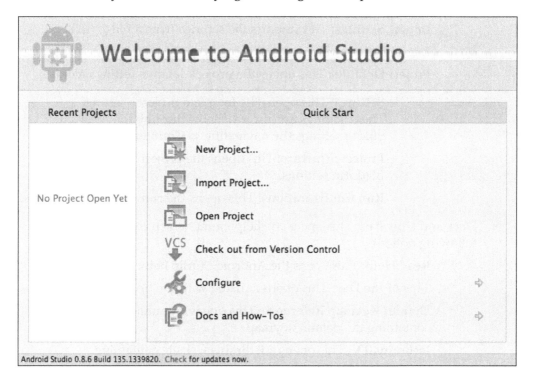

Let's take a look at the various options available in the **Quick Start** section:

- **Start a new Android Studio project**: This creates a new Android project from scratch.
- **Open an existing Android Studio Project**: This opens an existing project.
- **Import an Android code sample**: This imports a project containing Google code samples from GitHub.
- **Check out project from Version Control**: This creates a new project by importing existing sources from a version control system.

- **Import Non-Android Studio project**: This creates a new project by importing existing sources from your system.
- **Configure**: This opens the configuration menu. The configuration menu has the following options:
  - **SDK Manager**: This opens the Android SDK tool, which will be explained in *Chapter 6, Tools*.
  - **Settings**: This opens the Android Studio preferences.
  - **Plugins**: This opens the plugins manager for Android Studio.
  - **Import Settings**: This imports the settings from a file (.jar).
  - **Export Settings**: This exports the settings to a file (.jar).
  - **Project Defaults**: This opens the project defaults settings menu.

    Settings: This opens the template project's settings. These settings are also reachable through the Android Studio settings (by navigating to **Configure** | **Settings**).

    Project Structure: This opens the project and platform settings.

    Run Configurations: This opens the run and debug settings.
- **Docs and How-Tos**: This opens the help menu, which contains the following options:
  - **Read Help**: This opens the Android Studio help, an online version
  - **Tips of the Day**: This opens a dialog with the tip of the day
  - **Default Keymap Reference**: This opens an online PDF file containing the default keymap
  - **JetBrains TV**: This opens a JetBrains website containing video tutorials
  - **Plugin Development**: This opens a JetBrains website containing information for plugin developers

Configuring the Android SDK

The essential feature that needs to be configured correctly is the Android SDK. Although Android Studio automatically installs the latest Android SDK available, you should have everything you need beforehand to create your first application. It is important to check it and to learn how we can change it.

In the Android Studio welcome screen, navigate to **Configure | Project Defaults | Project Structure**. In **SDK Location**, you should have a selected **Android SDK location**, as shown in the next screenshot. This selected SDK location is the default location that will be used in our Android projects. However, we can change it later for specific projects that require special settings.

If you don't have any Android SDK configured in Android Studio, then you have to add it manually.

Installing and Configuring Android Studio

To accomplish this task, click on the ellipsis button to add an Android SDK, and then select the home directory for the SDK. Check whether you have it in your system by navigating to your Android Studio installation directory. You should find a folder named sdk. It contains the Android SDK and its tools. The Android Studio installation directory might be in a hidden folder, so click on the button highlighted in the following screenshot to show hidden files and directories:

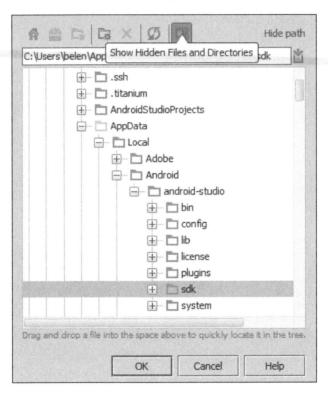

If you wish to use another Android SDK location that is different from the location included with Android Studio, select this instead. For example, if you previously used the **Android Development Tools** (**ADT**) plugin for Eclipse, you already have an Android SDK installed in your system.

Summary

We successfully prepared the system for Android Studio and installed our Android Studio instance. We ran the Studio for the first time, and now we know the options available on the welcome screen. Also, you learned how to configure your Android SDK and install it manually in case you wish to use a different version. Completing these tasks will leave our system with Android Studio running and configured to create our first project.

In the next chapter, we will learn about the concept of a project and how it includes everything the application requires, from classes to libraries. We will also create our first project and discuss the different kinds of activities available in the wizard.

2
Starting a Project

Now that you have installed Android Studio, the next thing to do is to get familiar with its features. You need to understand the necessary fields and form factors when creating a project. Also, you may need to choose the activity type to create the main activity. How can you achieve this using Android Studio?

In this chapter, we will discuss how to create a new project with the basic content that is needed to start out. We will use the Android Studio wizard to create the project and go through the project configuration fields. We will select one of the different kinds of activities available in the wizard as our main activity.

These are the topics we'll be covering in this chapter:

- Creating a new project
- Selecting the parameters
- Choosing your main activity from different types of activities

Starting a Project

Creating a new project

To create a new project, click on the **Start a new Android Studio project** option from the welcome screen. If you are not in the welcome screen, then navigate to **File | New Project**. This opens the **New Project** wizard, as shown in the following screenshot:

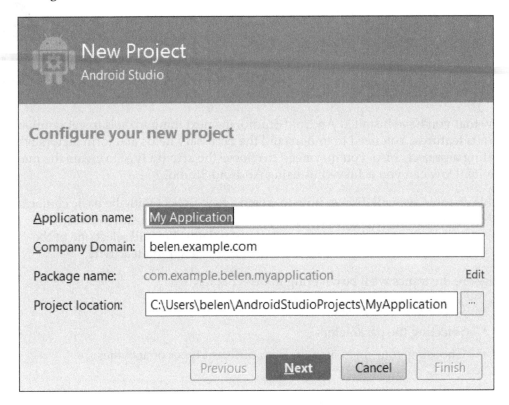

Configuring the project

The fields that will be shown in the **New Project** wizard are as follows:

- **Application name**: This is the name shown in Google Play and the name that users see.
- **Company Domain**: This is the company or domain name that is used to create the package name of your application.

- **Package name**: This is the unique identifier of your application, usually in the `com.company_name.app_name` or `reverse_company_domain.app_name` form. This form reduces the risk of name conflicts with other applications. A default package name is proposed based on the **Company Domain** and **Application name** fields. You can change the package name by clicking on **Edit**.
- **Project location**: This is the directory in your system in which the project is saved.

Complete the information for your project and click on the **Next** button. This will take you to the second screen. This screen allows you to select your platform and the minimum SDK your project will support on different devices.

Selecting the form factors

Because of the way Android has expanded to different types of devices, you may want to select one or more kinds of platforms and devices to be included in your project. For each type of device, you can select a different minimum SDK. The devices Android supports are as follows:

- **Phone and Tablet**: These are standard Android platforms used to create an application for phones and/or tablets
- **TV**: This is an Android TV platform used to design applications for big screens, such as those of television sets
- **Wear**: This is an Android Wear platform used to design applications for wearable devices such as smart watches
- **Glass**: This is an Android Glass platform used to design applications for Google Glass devices

To include any of these platforms in your project, you need to have them installed in your system. A tool known as Android SDK Manager has to be used to install a new platform. The Android SDK Manager tool will be explained in *Chapter 6, Tools*.

Once you've decided on your devices, you can choose the minimum SDK supported by your application. Devices with an older SDK will not be able to install your application. Try to reach a balance between supported devices and available features. If your application does not require a specific feature published in the newest SDKs, then you can select an older **application programming interface (API)**. The latest dashboards published by Google about platform distribution show that 99.5 percent of devices use Android 2.3 or later versions. If you select Android 2.2, then the percentage rises to 100 percent. You can check out these values by clicking on the **Help me choose** link. The official Android dashboards are also available at http://developer.android.com/about/dashboards/index.html.

Starting a Project

Check the **Phone and Tablet** option and select **API 15** as the minimum SDK. After that, click on **Next**. This will take you to the next screen, where you can select the activity type.

Choosing the activity type

Activities are the components associated with the screens with which users interact in an application. Android applications are usually based on multiple activities. When an application is launched, the activity indicated as the main activity is displayed. The third screen allows you to create the main activity of your application. You can also complete the creation of a new project without adding an activity.

Several types of activities that can be selected are as follows:

- **Blank Activity**: This creates a blank activity with an action bar. The action bar includes a title and an options menu. Action bars can provide navigation modes and user actions. You can read more about action bars at http://developer.android.com/guide/topics/ui/actionbar.html. The following screenshot shows **Blank Activity**:

- **Blank Activity with Fragment**: This creates a blank activity with an action bar and a contained **fragment**. A fragment is a portion of the user interface in an activity. Fragments can be reused in multiple activities, and multiple fragments can be combined in a single activity. You can find out more about fragments at https://developer.android.com/guide/components/fragments.html. Here is a screenshot showing **Blank Activity with Fragment**:

Chapter 2

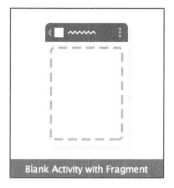

- **Fullscreen Activity**: This template hides the system user interface (such as the notification bar) in a fullscreen view. By default, the fullscreen mode is alternated with an action bar that shows up when the user touches the device screen. **Fullscreen Activity** is shown in the following screenshot:

- **Google Maps Activity**: This template creates a new activity with a Google map. It is shown in the next screenshot:

Starting a Project

- **Google Play Services Activity**: This template creates a new activity connected to the Google Play Services client. It is shown in the following screenshot:

- **Login activity**: This template creates a view as a login screen, allowing the users to log in or register with an e-mail and a password.

- **Master/Detail Flow**: This template splits the screen into two sections: a left-hand-side menu and the details of the selected item on the right-hand side. On a smaller screen, just one section is displayed, but on a bigger screen, both sections are displayed at the same time.

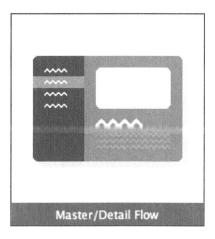

- **Navigation Drawer Activity**: This template creates a new activity with a navigation drawer. A navigation drawer displays the main navigation options in a panel that is brought onto the screen from a left-hand side panel. You can read more about navigation drawers at `https://developer.android.com/design/patterns/navigation-drawer.html`.

Starting a Project

- **Settings Activity**: This creates a preferences activity with a list of settings.

- **Tabbed Activity**: This creates a blank activity with an action bar similar to the **Blank Activity** menu, but it also includes a navigational element. The navigational element can be a tabbed user interface (fixed or scrollable tabs), a horizontal swipe, or a spinner menu.

Select **Blank Activity** and click on **Next**. In the final screen, we can give a name to the activity and its associated layout. Retain the default values and click on **Finish**.

Summary

We used the Android Studio wizard to create our first project, and filled the configuration fields. We also went through the different kinds of activities.

In the next chapter, we will go through the different elements of the structure of Android Studio. We will see how to create new classes, add and access libraries, and configure the project.

3
Navigating a Project

Now that you have created your first Android Studio project, you will understand what is going on. Before you start programming, you need to familiarize yourself with the navigation through the project. How is everything structured? Which settings can you change in the project? How can you change these settings and what do they mean?

This chapter is designed to introduce the structure of a project in Android Studio. We will start by examining the project navigation panel. Then we will go through the most important folders in our project—`build`, `gen`, and `libs`—and the folders under `src/main`, and you will learn how to change the project settings.

These are the topics we'll be covering in this chapter:

- The navigation panel
- The project structure
- Changing project properties

The project navigation panel

Initially, no project or file is displayed in the main view of Android Studio, as you can see in the next screenshot. As Android Studio suggests, press *Alt + 1* to open the project view. You can also open it by clicking on the **Project** button on the left edge of the screen.

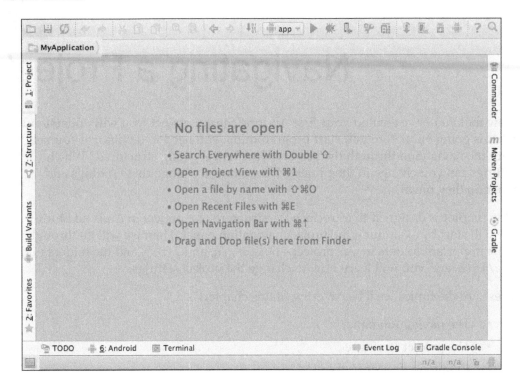

The **Project** view shows the list of open projects. These projects are displayed in a hierarchical view.

We can change the type of view to **Project**, **Packages**, or **Android** from the upper-left corner of the project explorer, as shown in the following screenshot. The **Project** view shows the directory structure of the project. The **Packages** view shows only the package structure. The **Android** view shows only the folders where you, as a developer, will include or edit your application files. These folders are related only to the Android application listed in a simplified structure: the `java` classes folder, the `res` resources folder, the `manifest` file, and the Gradle scripts.

In the upper-right corner of the screen, there are some actions and a drop-down menu to configure the **Project** view. These actions are highlighted in the following screenshot:

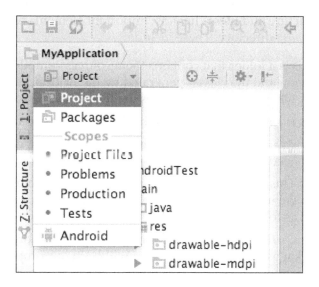

Right-click on the project name to open the context menu, or click on any element of the project. From this menu, we can:

- Create and add new elements to the project
- Cut, copy, paste, or rename files in the project
- Find the elements in the project
- Analyze and reformat the code
- Build the project
- Compare files
- Open files in Explorer

Navigating a Project

The project structure

We can examine the project structure in the project navigation pane. The project structure includes a folder with the name of our application. This folder contains the application structure and files. The most important elements of the application structure are in the app directory. These include:

- build/: This is a folder that contains the resources compiled after building the application and the classes generated by the Android tools, such as the R.java file, which contains references to the application resources.
- libs/: This is a folder that contains the libraries referenced from our code.
- src/androidTest/: This is a folder that contains the test classes of the Java classes that need to be tested.
- src/main/: This is a folder that contains the sources of our application. All the files we usually work with are in this folder. The main folder is subdivided as follows:
    - java/: This is a folder that contains Java classes organized as packages. Every class we create will be in our project package namespace (com.example.myapplication). When we created our first project, we also created its main activity, so the activity class should be in this package. The following screenshot shows this main activity class inside the project structure:

- ○ `res/`: This is a folder that contains project resources such as the XML files that specify layouts and menus, or image files.
  - ○ `AndroidManifest.xml`: This is an essential file in an Android project, which is generated automatically when we create the project. This file declares the basic information needed by the Android system to run the application: package name, version, activities, permissions, intents, or required hardware.
- `build.gradle`: This file is the script used to build our application. We will discuss how to configure options in this file in the *Gradle* subsection.

The resources folder

The resources are all non-code assets associated with our application. Elements such as images or strings are externalized from the code as resources, making it easy to update them without changing the code. Some examples of resources include colors, images, graphics, layouts, strings, and styles. The resources are distributed in the following folders:

- `color/`: This is a folder that contains the color state lists used in our application. The color state lists define colors and color changes based on the component states.
- `drawable/`: This is a folder that contains the images used in our application. There are different drawable folders categorized into different screen densities. When we created our first project, a default application icon was also created. This icon, named `ic_launcher.png`, is already in these folders.
- `layout/`: This is a folder that contains the XML definitions of the views and their elements.
- `menu/`: This is a folder that contains the XML definitions of the menus of the application.
- `values/`: This is a folder that contains the XML files that define sets of name-value pairs. These values can be colors, strings, or styles. There are different `values` folders categorized into different screen options to adapt the interface to them; for example, to enlarge the components or the fonts when the application is running on a tablet.

Our basic project contains some basic resources. Therefore, all the folders discussed here are not necessarily included by default.

Gradle

Applications in Android Studio are built using Gradle. Gradle is a build automation tool that is independent of Android Studio but totally integrated with it. Gradle uses an extensible and declarative **domain-specific language** (**DSL**) that is based on Groovy. A Gradle build file consists of one or more projects, and each project contains one or more tasks. A task represents a piece of work to be built. You can learn more about Gradle at http://www.gradle.org/.

The configuration for the build process is declared in the Gradle build files included in the Android projects. As explained previously, in the project structure, the build configuration file of the Android application is defined in the /app/build.gradle file. Some of the main options we can configure in this file are as follows:

- **Variants**: We can configure different versions of our application using the same project, for example, to create demo and paid versions. The variants depend on the build type (the buildTypes tag) and product flavor configurations (the productFlavors tag). For example, two build types are debug and release, and two product flavors are demo and paid.

- **Dependencies**: We can indicate the local or remote dependencies of our project on other modules or libraries. These dependencies are declared under the dependencies tag.

- **Manifest entries**: We can override some entries of the manifest Android file in the build file, providing a dynamic configuration of the manifest file. For example, we can override the values of the package name, the minimum SDK, or the target SDK. These configurations are defined under the android/defaultConfig tags.

- **Signing**: We can activate the application signing for the release version. The build system uses a default certificate to sign the debug version of the application. We can configure our key and certificate to sign the release version as well. These configurations are defined under the android/signingConfigs tags.

Project settings

You can navigate to the two dialogs that contain project settings using the following: **File | Settings** and **File | Project Structure**. Both are also available in the toolbar.

Select your project from the project view and navigate to the **Settings** menu in **File**. The left-hand-side panel of the **Settings** dialog displays a section named **Project Settings [MyApplication]**. Some important options are as follows:

- **Code Style**: This configures the default code style scheme.
- **Compiler**: This configures the Android DX compiler used when building our application.
- **File Encodings**: This changes the file's encoding. The default encoding is UTF-8.
- **Gradle**: This gives Gradle's configuration.
- **Language Injections**: This adds or removes the available languages used in the editor.
- **Version Control**: This configures the version control options. Version control will be explained in more detail in *Chapter 6, Tools*.

In addition to these, there are further settings in the **Project Structure** dialog. Navigate to **File | Project Structure**. The settings include the following:

- **SDK Location**: We can change the project SDK. In *Chapter 1, Installing and Configuring Android Studio*, we selected an SDK as the default. In this screen, we can change this SDK, just for the current project.
- **Project**: We can change the Gradle version or the plugin and library repository.
- **Modules**: According to IntelliJ IDEA (http://www.jetbrains.com/idea/webhelp/module.html), the following is the definition of a module:

    A module is a discrete unit of functionality which you can compile, run, test and debug independently.

    The **Modules** menu shows a list of existing modules with their facets. The default module we have in our project is the `app` module. The settings tabs correspond to the following Gradle build file configurations: **Properties**, **Signing**, **Flavors**, **Build Types**, and **Dependencies**.

- **Libraries**: This menu shows a list of the libraries imported to the project. We can also remove them or add new libraries. They will be added to the `libs/` folder.

Summary

We saw how projects are presented in Android Studio and which folders they contain by default when created. We explored the reasons for having those folders and examined the `AndroidManifest.xml` file and its purpose. We also went through the project settings in the **Preferences** and **Project Structure** dialogs. By now, you know how to manipulate and navigate through a project in Android Studio.

In the next chapter, we will discuss how to use the text editor. Proper knowledge of the text editor is important in order to improve our programming efficiency. Next, we will learn about the editor settings and how to autocomplete code, use pregenerated blocks of code, and navigate through the code. You will also learn about some useful shortcuts.

4
Using the Code Editor

Now that you have created your first project and learned how to navigate through different folders, subfolders, and files, it's time to start programming. Have you ever wanted to be able to program more efficiently? How can you speed up your development process? Do you want to learn useful shortcuts too? For example, how can you comment more than one line at once, find and replace strings, or move faster through different parameters in a method call?

In this chapter, you will learn how to use and customize the code editor in order to feel more comfortable when programming. It is worth learning the basic features of the code editor in order to increase your productivity. You will learn about code completion and code generation. Finally, you will learn some useful shortcuts and hotkeys to speed up the development process.

These are the topics we'll be covering in this chapter:

- Customizing the code editor
- Code completion
- Code generation
- Finding related content
- Useful shortcuts

Using the Code Editor

Customizing the editor settings

To open the editor settings, navigate to **File | Settings**. In the **IDE Settings** section of the left panel, select **Editor**. This displays the general settings of the editor in the right panel. We recommend checking two of the options that are unchecked by default:

- **Change font size (Zoom) with Ctrl+Mouse Wheel**: Checking this option allows us to change the font size of the editor using the mouse wheel, as we do in other programs such as web browsers.

- **Show quick doc on mouse move**: Checking this option enables the display of a brief document about the code in a small dialog when we move the mouse over a piece of code and wait for 500 milliseconds. When we move the mouse again, the dialog automatically disappears, but if we move the mouse into the dialog, then we can examine the document in detail. This is very useful to read what a method does and to identify its parameters without navigating to it. The following screenshot displays this functionality:

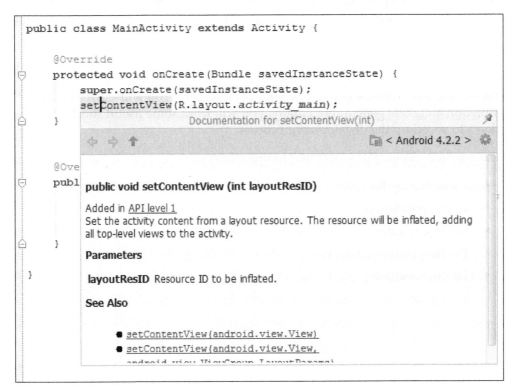

There are more settings, distributed among seven categories as follows:

- **Smart Keys**: This category configures actions to be done automatically while typing, such as adding closing brackets, quotes, or tags, and indenting the line when we press the *Enter* key.
- **Appearance**: This category configures the appearance of the editor. Here, you can change the theme, customize the fonts and colors, and so on. We recommend checking the next two options. They are unchecked by default:
    - **Show line numbers**: This shows the line numbers on the left edge of the editor. It can be very useful when we are debugging or examining the log.
    - **Show method separators**: This visually separates the methods of a class.
- **Colors & Fonts**: This category changes fonts and colors. There are a lot of options and elements to configure (keywords, numbers, warnings, errors, comments, strings, and so on). We can save the configurations as schemes.
- **Editor Tabs**: This category configures the editor tabs. We advise you to select the **Mark modified tabs with asterisk** option to easily detect modified and unsaved files.
- **Code Folding**: This category collapses or expands code blocks. The **code folding** option allows us to hide code blocks that we are not editing, simplifying the code view. We can collapse or expand the blocks using the icons from the editor, as shown in the following screenshot, or by using the **Folding** menu from **Code**:

```
1    package com.example.myapplication;
2
3    import ...
6
7    public class MainActivity extends Activity {
8
9        @Override
10       protected void onCreate(Bundle savedInstanceState) {
11           super.onCreate(savedInstanceState);
12           setContentView(R.layout.activity_main);
13       }
14
```

- **Code Completion**: This category configures the code completion options. We will examine code completion in detail in the next section.
- **Auto Import**: This category configures how the editor behaves when we paste code that uses classes not imported into the current class. By default, when we do this, a pop-up appears, and it tells us to add the `import` command. If we check the **Add unambiguous imports on the fly** option, the `import` command is added automatically, without our interaction.

```
1   package com.example.myapplication;
2
3   import android.os.Bundle;
4   import android.app.Activity;
5   import android.view.Menu;
6
7   public class MainActivity extends Activity {
8
9       @Override
10      protected void onCreate(Bundle savedInstanceState) {
11          ? android.util.Log? Alt+Enter  ate(savedInstanceState);
12                         setContentView(R.layout.activity_main);
13          Log.i("MainActivity", "Test");
14      }
```

Code completion

Code completion helps us write code quickly by automatically completing the code using dynamic suggestion lists that are generated based on what we just typed.

The basic code completion is the list of suggestions that appears while we are typing, as shown in the following screenshot. If the list is not displayed, press *Ctrl* and the spacebar to open it.

```
 9        @Override
10        protected void onCreate(Bundle savedInstanceState) {
11            super.onCreate(savedInstanceState);
12            setContentView(R.layout.activity_main);
13
14            L
15        ┌─ LAYOUT_INFLATER_SERVICE                          String
16        │  LOCATION_SERVICE                                 String
17        │  LinkageError   (java.lang)
18        │  Long           (java.lang)
19        │  databaseList ()                                  String[]
20        │  fileList ()                                      String[]
21        │  getClassLoader ()                                ClassLoader
22        │  getLayoutInflater ()                             LayoutInflater
23        │  getLoaderManager ()                              LoaderManager
24        │  getLocalClassName ()                             String
25    }
26        Press Ctrl+Punto to choose the selected (or first) suggestion and insert a dot afterwards >>
```

Keep typing, select a command from the list, and press *Enter* or double-click to add it to your code. If the code you are writing is an expression and you want to insert the expression in its negated form, then select the expression from the suggestion list, and instead of pressing *Enter* or double-clicking it, press the exclamation mark key (*!*). The expression will be added with negation.

Another type of code completion is **smart type** code completion. If you are typing a command to call a method with a `String` parameter, then only the `String` objects will be suggested. This smart completion occurs in the right-hand side of an assignment statement, parameters of a method call, return statements, or variable initializers. To open the smart suggestions list, press *Ctrl + Shift* along with the spacebar.

Using the Code Editor

To show the difference between these two types of suggestion lists, create two objects of different classes, String and int, in your code. Then call a method with a String parameter, for example, the i method of the Log class. When typing the String parameter, note the difference between opening the basic suggestion list (*Ctrl + spacebar*), which the next screenshot shows, and opening the smart type suggestion list (*Ctrl + Shift + spacebar*), which the second screenshot shows.

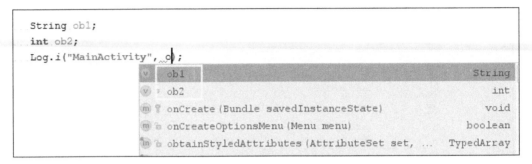

In the first list, which is shown in the previous screenshot, both objects are suggested, although the int object does not match the parameter class. In the second list, which is shown in the following screenshot, only String objects are suggested:

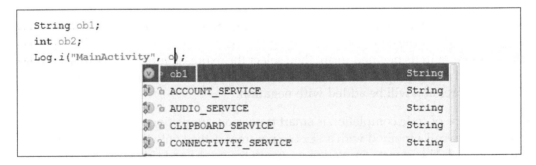

One last utility of code completion is the **completion of statements**. Type a statement, press *Ctrl + Shift + Enter*, and notice how the closing punctuation is automatically added. If you press these keys after typing the if keyword, then the parentheses and the brackets are added to complete the conditional statement. This shortcut can also be used to complete method declarations. Start typing a method, and after typing the opening parenthesis or the method parameters, press *Ctrl + Shift + Enter*. The closing parenthesis and the brackets are added to complete the method specification.

Code generation

To generate blocks of code in a class, navigate to **Code | Generate** or use the *Alt + Insert* shortcut. We can generate constructors, getters and setters methods, and `equals` and `toString` methods. We can also override or delegate methods.

Another way of generating code is surrounding some of our code with statements (`if`, `if/else`, `while`, `for`, `try/catch`, and so on). Select a code line and navigate to **Code | Surround With** or press *Ctrl + Alt + T*.

The third option is inserting code templates. Navigate to **Code | Insert Live Templates** or press *Ctrl + J* to open a dialog of the available templates. These templates can insert code to iterate collections, arrays, lists, and so on; code to print formatted strings; code to throw exceptions; or code to add static and final variables. The left edge of the dialog shows the prefix for each template. If you type the prefix in the editor and press the *Tab* key, the code template is added automatically.

Type `inn` at the end of the `onCreate` method of our main activity and press *Tab*. A conditional block will appear. In this new block, type `soutv` and press *Tab* again. The result is as follows:

```
protected void onCreate(Bundle savedInstanceState) {
    super.onCreate(savedInstanceState);
    setContentView(R.layout.activity_main);

    if (savedInstanceState != null) {
        System.out.println("savedInstanceState = " + savedInstanceState);
    }
}
```

Navigating code

The most direct way of navigating to declarations or type declarations is by pressing *Ctrl* and clicking on the method name when it is displayed as a link. This option is also accessible from the **Declaration** menu in **Navigate**.

Using the Code Editor

We can navigate through the hierarchy of methods from the left edge of the editor. Next to the method declarations that belong to a hierarchy of methods, there is an icon that indicates whether a method is implementing an interface method, implementing an abstract class method, overriding a superclass method, or getting implemented or overridden by other descendants. Click on these icons to navigate to the methods in the hierarchy. This option is also available via **Navigate | Super Method** or **Navigate | Implementation(s)**. You can test it in the main activity of our first project (`MainActivity.java`), as shown in the following screenshot:

```
 7  public class MainActivity extends Activity {
 8
 9      @Override
10  ⮑  Overrides method in 'android.app.Activity' ndle savedInstanceState) {
11          super.onCreate(savedInstanceState);
12          setContentView(R.layout.activity_main);
13      }
14
```

Another useful utility related to code navigation is the use of **custom regions**. A custom region is a piece of code that you want to group and name. For example, if there is a class with many methods, we can create custom regions to distribute the methods among them. A region has a name or description, and it can be collapsed or expanded using code folding.

To create a custom region, we can use code generation. Select the fragment of code, navigate to **Code | Surround With**, and select one of these two options:

- **<editor-fold...> Comments**
- **region...endregion Comments**

Both of these options create a region but use different styles.

When we are using custom regions, we can navigate to them using the **Custom Region** menu in **Navigate**. The rest of the navigation options are accessible from the **Navigate** menu. Some of these options are as follows:

- **Class/File/Symbol**: This finds a class, file, or symbol by its name.
- **Line**: This option goes to a line of code by its number.
- **Last Edit Location**: This navigates to the most recent change point.

- **Test**: This navigates to the test of the current class.
- **File Structure**: This opens a dialog that shows the file structure. Open the file structure of our main activity and observe how the structure is presented, displaying the list of methods and the icons that indicate the type or visibility of the element, as shown in the following screenshot:

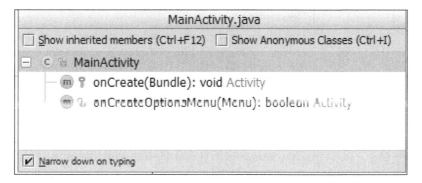

- **File Path**: This opens a dialog that shows the complete path to the file opened in the editor.
- **Type Hierarchy**: This opens a dialog that shows the type hierarchy of the selected object.
- **Method Hierarchy**: This opens a dialog that shows the method hierarchy of the selected method.
- **Call Hierarchy**: This opens a dialog that shows the call hierarchy of the selected method.
- **Next Highlighted Error**: This navigates to the next error.
- **Previous Highlighted Error**: This navigates to the previous error.
- **Next Method**: This navigates to the next method.
- **Previous Method**: This navigates to the previous method.

Useful shortcuts

You can find all the available shortcuts and change them through the **Keymap** option in the **IDE Settings** section of **Settings**. Some useful shortcuts for Windows are included in the following list:

Shortcut	Description
Ctrl + W	This selects expressions based on grammar. Press these keys repeatedly to expand the selection. The opposite command is Ctrl + Shift + W.
Ctrl + /	This comments each line of the selected code. To block comments, use Ctrl + Shift + /.
Ctrl + Alt + I	This indents the selected code. This is useful when cleaning up a block of code or method after you finish writing.
Ctrl + Alt + O	This optimizes the imports, removing the unused imports and reordering the rest of them.
Shift + Ctrl + Arrows	This moves the selected code a line above or below.
Alt + Arrows	This switches between the opened tabs of the editor.
Ctrl + F	This finds a string in the active tab of the editor.
Ctrl + R	This replaces a string in the active tab of the editor.
Ctrl + A	This selects all of the code of the opened file.
Ctrl + D	This copies the selected code and pastes it at the end of the selection. If no code is selected, then the entire line is copied and pasted in a new line.
Ctrl + Y	This removes the entire line without leaving a blank line.
Ctrl + Shift + U	This toggles the case.
Tab	This moves to the next parameter.

Summary

By the end of this chapter, you should have learned some useful tricks and tips to make the most of the code editor. You now know how to use code completion, code generation, and some useful shortcuts to speed up different actions. We customized our code editor and are now ready to start programming.

In the next chapter, we will start creating our first user interface using layouts. You will learn how to create a layout using the graphical wizard, as well as by editing the XML layout file using the text-based view. We will create our first application, a classic `Hello World` example, using the text view component. You will learn how to prepare the application for multiple screen sizes and adapt them for different device orientations. Finally, you will learn about UI themes and how to handle events.

5
Creating User Interfaces

Now that we have created our first project and have become familiar with the code editor and its functionalities, we will begin our application by creating our user interface. Is there more than one way to create a user interface using Android Studio? How can you add components to your user interface? Have you ever wondered how to make your applications support different screen sizes and resolutions?

This chapter focuses on the creation of user interfaces using layouts. Layouts can be created using a graphical view or a text-based view. You will learn how to use both of them to create our layout. We will also code a `Hello World` application using simple components. Since there are over 18,000 Android devices, you will learn about fragmentation on different Android-based devices and will discuss how to prepare our application for this issue. We will end this chapter with basic notions of handling events on our application.

These are the topics we'll be covering in this chapter:

- Existing layout editors
- Creating a new layout
- Adding components
- Supporting different screens
- Changing the UI theme
- Handling events

Creating User Interfaces

The graphical editor

Open the main layout located at `/src/main/res/layout/activity_main.xml` in our project. The graphical editor will be opened by default. Initially, this main layout contains just a text view with a **Hello world!** message. To switch between the graphical and the text editor, click on the **Design** and **Text** tabs at the bottom of the screen, as shown in this screenshot:

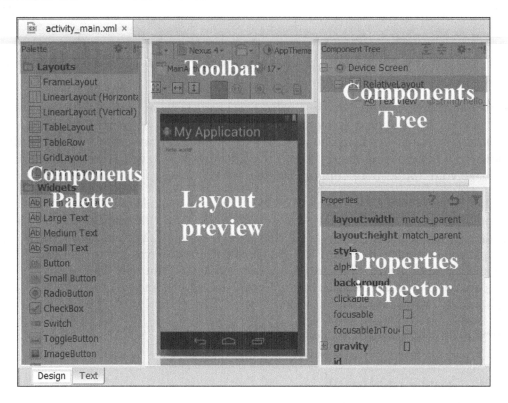

Toolbar contains some options that can be used to change the layout style and preview. The **Toolbar** options, which are shown in the following screenshot, are explained throughout the chapter:

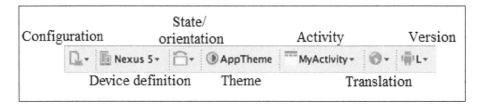

Components Tree displays the components placed in the layout as a hierarchy. **Properties inspector** shows the properties of the selected component from the layout and allows us to change them. **Components Palette** lists the existing **user interface** (**UI**) components to place in our layout. It organizes the components in different categories. Let's look at the options available in **Components Palette**:

- **Layouts**: A layout is a container object used to distribute the components on the screen. The root element of a UI is a layout object, but layouts can also contain more layouts, creating a hierarchy of components structured in layouts. The recommendation is to keep this layout hierarchy as simple as possible. Our main layout has a **relative layout** as the root element.

- **Widgets**: This category contains options for text views, buttons, checkboxes, switches, image views, progress bars, spinners, and web views. They are the most commonly used components, and they are used in most layouts.

- **Text Fields**: These are editable fields that contain different categories of inputs under which users can type text. The difference between the various options is the type of text users can type.

- **Containers**: This category groups components that share some common behavior. Radio groups, list views, scroll views, and tab hosts are some of them.

- **Date & Time**: This category holds components related to date and time in the form of calendars or clocks.

- **Expert**: The components in this category are not as common as the components in the **Widgets** category, but it is worth taking a look at them.

- **Custom**: This category holds components that allow us to include our custom components, which are usually other layouts from our project.

The text-based editor

Change the graphical editor to the text editor by clicking on the **Text** tab.

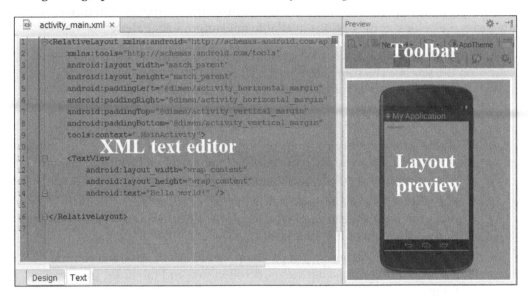

The **Toolbar** is the same as that on the graphical editor. The **Preview** window displays the layout but it cannot be changed. To do that, you should use the **Design** tab instead. The components are added to the layout using their XML declarations. The properties are also configured using XML declarations. Like the graphical editor, the text editor shows just the text view element inside the root layout.

Creating a new layout

When we created our main activity, the associated layout was also created. This is a way of creating a layout when creating an activity.

To add an independent layout without creating a new activity, right-click on the `layouts` folder (`res/layout/`) and navigate to **New | Layout resource file**. You can also navigate to this menu option like this: **File | New | Layout resource file**. Type the file name and the root element.

Once the layout is created, the associated activity can be changed from the editor to another one. If the layout has no activity, any existing activity can be linked to it from the editor. To accomplish this, search for the **Associate with Activity** option in the toolbar of the layout editor, click on it, and select the **Associate with other Activity** option. A dialog box that lists all the activities of your project will open, and you can select one of them.

Adding components

Our main layout is a relative layout and contains a text view saying **Hello world!**. Now let's add a new component. The easiest way to do this is by using the graphical editor, so open the **Design** tab. Select a component and drag it into the layout preview; for example, navigate to the **Person Name** component in **Text Fields** and place it below the text view.

In the **Component Tree** view, there is now a new `editText` object. Keep the text field selected to examine its properties loaded in the **Properties** inspector. Let's change some of them and observe the differences in the layout preview and in the component tree:

- **layout:width**: This option will adapt the width of the field to its content. Its current value is `wrap_content`. Change it to `match_parent` to adapt it to the parent layout width (the root relative layout).
- **hint**: Type `Enter your name` as the hint of the field. The hint is a text shown when the field is empty to indicate the information that should be typed. As the field has a default value, `Name`, the hint is not visible.
- **id**: This has `@+id/editText` as the current ID. This ID will be used from the code to get access to this object and is the ID displayed in the component tree. Change it to `@+id/editText_name` to distinguish it easily from other text fields. Check whether the component ID has also changed in the **Component Tree** window, as shown in the following screenshot:

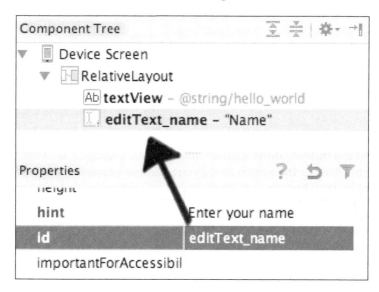

- **text**: This deletes the value of this field. The `hint` value should now be visible.

If we switch to the text editor, we can see the XML definition of the text field with the properties we edited:

```
<EditText
  android:layout_width="match_parent"
  android:layout_height="wrap_content"
  android:inputType="textPersonName"
  android:ems="10"
  android:id="@+id/editText_name"
  android:layout_below="@+id/textView_greeting"
  android:layout_alignLeft="@+id/textView_greeting"
  android:layout_marginTop="15dp"
  android:hint="Enter your name"
/>
```

From the text editor, the existing components and their properties can also be changed. Modify the text view ID (the `android:id` property) from `@+id/textView` to `@+id/textView_greeting`. Having a descriptive ID is important because it will be used by our code. Descriptive variable names allow the code to be self-documenting.

Let's add another component using the text editor this time. Press the open tag key and start typing `Button`. Let the list of suggestions appear and select a `Button` object. Inside the `Button` tag, add the following properties:

```
<Button
  android:id="@+id/button_accept"
  android:layout_width="wrap_content"
  android:layout_height="wrap_content"
  android:layout_below="@+id/editText_name"
  android:layout_centerHorizontal="true"
  android:text="Accept"
/>
```

Create the ID property with the value of `@+id/button_accept`. Let the width and height adapt to the button content (the `wrap_content` value). Place the button below the name text field using the `android:layout_below` property. We reference the name text field by its ID (`@+id/editText_name`). Center the button horizontally in the parent layout using the `layout_centerHorizontal` property. Set the text of the button (`Accept`).

The button is displayed in **Preview**. The next screenshot shows that if we switch to the graphical editor, the button is displayed in it and also in **Component Tree**:

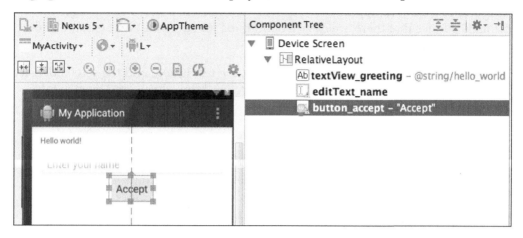

Supporting multiple screens

When creating Android applications, we have to take into account the existence of multiple screen sizes and screen resolutions. It is important to check how our layouts are displayed in different screen configurations. To accomplish this, Android Studio provides a functionality to change the virtual device that renders the layout preview when we are in the **Design** mode.

We can find this functionality in the toolbar and click on it to open the list of available device definitions, as shown in the following screenshot:

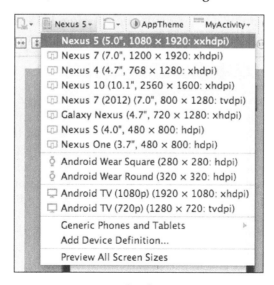

Creating User Interfaces

Try some of them. The difference between a tablet device and a device like those from the Nexus line is very notable. We should adapt the views to all the screen configurations that our application supports to ensure that they are displayed optimally. Notice that there are device definitions for Android Wear (square and round designs) and for Android TV.

The device definitions indicate the screen size, the resolution, and the screen density. Android screen densities include ldpi, mdpi, tvdpi, hdpi, xhdpi, and even xxhdpi. Let's see what their values are:

- **ldpi**: This is **low-density dots per inch**, and its value is about 120 dpi
- **mdpi**: This is **medium-density dots per inch**, and its value is about 160 dpi
- **tvdpi**: This is **medium high density dots per inch**, and its value is about 213 dpi
- **hdpi**: This is **high-density dots per inch**, and its value is about 240 dpi
- **xhdpi**: This is **extra-high-density dots per inch**, and its value is about 320 dpi
- **xxhdpi**: This is **extra-extra-high-density dots per inch**, and its value is about 480 dpi

The latest dashboards published by Google show that most devices have high-density screens (36.4 percent), followed by mdpi (19.6 percent) and xhdpi (19.3 percent). Therefore, we can cover 75.3 percent of all devices by testing our application using these three screen densities. If you want to cover a bigger percentage of devices, test your application using xxhdpi screens (15.2 percent) as well so that the coverage will be 90.5 percent of all devices. The official Android dashboards are available at `http://developer.android.com/about/dashboards`.

Another issue to keep in mind is the **device orientation**. Do we want to support landscape mode in our application? If the answer is yes, then we have to test our layouts in landscape orientation. On the toolbar, click on the **layout state** option to change the mode either from portrait to landscape or from landscape to portrait.

If our application supports landscape mode and the layout does not get displayed as expected in this orientation, we might want to create a variation of the layout. Click on the first icon of the toolbar, that is, the **Configuration to render this layout with inside the IDE** option, and select the **Create Landscape Variation** option. A new layout will be opened in the editor. This layout has been created in the `resources` folder, under the `layout-land` directory, and it uses the same name as the portrait layout—`/src/main/res/layout-land/activity_main.xml`. Now we can edit the new layout variation such that it perfectly conforms to landscape mode.

Similarly, we can create a variation of the layout for extra-large screens. Select the **Create layout-xlarge Variation** option. The new layout will be created in the layout-xlarge folder at /src/main/res/layout-xlarge/activity_main.xml. Android divides into the actual screen sizes small, normal, large, and xlarge:

- **Small**: Screens classified in this category are at least 426 dp x 320 dp
- **Normal**: Screens classified in this category are at least 470 dp x 320 dp
- **Large**: Screens classified in this category are at least 640 dp x 480 dp
- **Xlarge**: Screens classified in this category are at least 960 dp x 720 dp

A **dp** is a **density-independent pixel**, equivalent to one physical pixel on a 160 dpi screen. The last dashboards published by Google show that most devices have a normal screen size (80.9 percent).

To display multiple device configurations at the same time, click on the **Configuration to render this layout with inside the IDE** option in the toolbar and select the **Preview All Screen Sizes** option, or click on the **Preview Representative Sample** option to open only the most important screen sizes, as shown in the following screenshot. We can also delete any of the samples by right-clicking on them and selecting the **Delete** option from the menu. Another useful action of this menu is the **Save screenshot** option. It allows us to take a screenshot of the layout preview.

If we create different layout variations, we can preview all of them by selecting the **Preview Layout Versions** option.

Now that we have seen how to add different components and optimize our layout for different screens, let's start working with themes.

Changing the UI theme

Layouts and widgets are created using the default UI theme of our project. We can change the appearance of the elements of the UI by creating styles. Styles can be grouped to create a theme, and a theme can be applied to an activity or to the whole application. Some themes are provided by default, such as the **Holo** style. Styles and themes are created as resources under the `/src/res/values` folder.

To continue our example, follow these steps:

1. Open the main layout using the graphical editor. The selected theme for our layout is indicated as **AppTheme** in the toolbar. This theme was created for our project and can be found in the `styles` file at `/src/res/values/styles.xml`.

2. Open the `styles` file. You will notice that this theme is an extension of another theme, `Theme.AppCompat.Light.DarkActionBar`.

3. To customize the theme, edit the `styles` file. Add the highlighted line in the **AppTheme** definition to change the window background color:

   ```
   <style name="AppTheme"
     parent="android:Theme.AppCompat.Light.DarkActionBar">
     <item name="android:windowBackground">
       @color/custom_theme_color</item>
   </style>
   <color name="custom_theme_color">#dddddd</color>
   ```

4. Save the file and switch to the layout tab. The background is now light gray. This background color will be applied to all our layouts due to the fact that we configured it in the theme and not just in the layout.

5. To change the layout theme completely, click on the theme option from the toolbar in the graphical editor. The theme selector dialog is now opened, displaying a list of the available themes, as shown in the following screenshot:

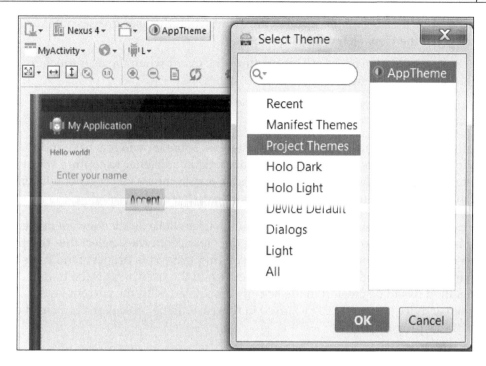

The themes created in our own project are listed in the **Project Themes** section. The **Manifest Themes** section shows the theme configured in the application manifest file (/src/main/AndroidManifest.xml). The **All** section lists all the available themes.

Handling events

The user interface would be useless if the rest of the application could not interact with it. Events in Android are generated when the user interacts with our application. All the UI widgets are children of the View class, and they share some events handled by the following listeners:

- OnClickListener: This captures the event when the user clicks on the view element
- OnCreateContextMenu: This captures the event when the user performs a long click on the view element and we want to open a context menu
- OnDragListener: This captures the event when the user drags and drops the event element
- OnFocusChange: This captures the event when the user navigates from one element to another in the same view

- `OnKeyListener`: This captures the event when the user presses any key while the view element has the focus
- `OnLongClickListener`: This captures the event when the user touches the view element and holds it
- `OnTouchListener`: This captures the event when the user touches the view element

In addition to these standard events and listeners, some UI widgets have more specific events and listeners. Checkboxes can register a listener to capture when its state changes (`OnCheckedChangeListener`), and spinners can register a listener to capture when an item is clicked (`OnItemClickListener`).

The most common event to capture is when the user clicks on the view elements. There is an easy way to handle this—using the view properties. Select the **Accept** button in our layout and look for the `onClick` property. This property indicates the name of the method that will be executed when the user presses the button. This method has to be created in the activity associated with the current layout, our main activity (`MainActivity.java`) in this case. Type `onAcceptClick` as the value of this property.

Open the main activity to create the method definition. When a view is clicked, the event callback method has to be public with a `void` return type. It receives the view that has been clicked on as a parameter. This method will be executed every time the user clicks on the button:

```
public void onAcceptClick(View v) {
  // Action when the button is pressed
}
```

From the main activity, we can interact with all the components of the interface, so when the user presses the **Accept** button, our code can read the text from the name field and change the greeting to include the name in it.

To get the reference to a `view` object, use the `findViewById` method inherited from the `Activity` class. This method receives the ID of the component and returns the `View` object corresponding to that ID. The returned `view` object has to be cast to its specific class in order to use its methods, such as the `getText` method of the `EditText` class, to get the name typed by the user:

```
public void onAcceptClick(View v) {
  TextView tv_greeting =
    (TextView) findViewById(R.id.textView_greeting);
  EditText et_name = (EditText) findViewById(R.id.editText_name);

  if(et_name.getText().length() > 0) {
```

```
      tv_greeting.setText("Hello " + et_name.getText());
    }
}
```

In the first two lines of the method, the following references to the elements of the layout are retrieved: the text view that contains the greeting, and the text field where the user can type a name. The components are found by their IDs, the same ID that we indicated in the properties of the element in the layout file. All the IDs of resources are included in the R class. The R class is autogenerated in the build phase, and therefore, we must not edit it. If this class is not autogenerated, then probably some file of our resources contain an error.

The next line is a conditional statement used to check whether the user typed a name. If they typed a name, the text will be replaced by a new greeting that contains that name. In the coming chapters, we will learn how to execute our application in an emulator, and we will be able to test this code.

If the event we want to handle is not the user's click, then we have to create and add the listener by code to the onCreate method of the activity. There are two ways to do this:

- Implementing the listener interface in the activity and then adding the unimplemented methods. The methods required by the interface are the methods used to receive the events.
- Creating a private anonymous implementation of the listener in the activity file. The methods that receive the events are implemented in this object.

Finally, the listener implementation has to be assigned to the view element using the setter methods, such as setOnClickListener, setOnCreateContextMenu, setOnDragListener, setOnFocusChange, setOnKeyListener, and so forth. The listener assignment is usually included in the onCreate method of the activity. If the listener is implemented in the same activity, then the parameter indicated to the setter method is its own activity using the this keyword, as shown in the following code:

```
Button b_accept = (Button) findViewById(R.id.button_accept);
b_accept.setOnClickListener(this);
```

The activity should then implement the listener and the onClick method required by the listener interface:

```
public class MyActivity extends Activity
implements View.OnClickListener {
  @Override
  public void onClick(View view) {
    // Action when the button is pressed
  }
```

Summary

In this chapter, we saw how to create and edit the user interface layouts using both the graphical and the text-based editors. We had our first small application finished, and we upgraded it with some basic components. You should now be able to create a simple layout and test it with different styles, screen sizes, and screen resolutions. You also learned about the different available UI themes. Finally, you learned about events and learned how to handle them using listeners.

In the next chapter, you will learn about the available Google Play services and how to integrate them with your project using Android Studio. We will also see how to install and integrate different libraries available with Google technology, such as Google Maps, Google Plus, and more.

6
Tools

In the previous chapter, you learned about the useful services that Google provides, which can be used by developers to improve their applications. Now you will learn about the tools available in Android Studio that make your life easier. Have you wondered how to manage Android platforms? Do you want to have your project clearly documented? Are you working as a group of developers and need a version control manager integrated with Android Studio?

This chapter describes the most important additional tools provided in Android Studio: Android SDK tools, Javadoc, and version control integration. First, you will learn about the **software development kit** (**SDK**) Manager available in Android Studio from which you'll be able to examine, update, and install different components for your project. Next, we will review the **Android Virtual Device** (**AVD**) Manager, where we can edit the virtual devices in which we will be testing our project. You will also learn how to have complete documentation using the Javadoc tool, and how to have version control using the systems available in Android Studio.

These are the topics we'll be covering in this chapter:

- SDK Manager
- AVD Manager
- Navigation Editor
- Javadoc
- Version control

The SDK Manager

The SDK Manager is an Android tool accessible from Android Studio to control our Android SDK installation. From this tool, we can examine the Android platforms installed in our system, update them, install new platforms, or install some other components such as Google Play services or Android Support Library.

To open the SDK Manager from Android Studio, navigate to **Tools** | **Android** | **SDK Manager**. You can also click on the shortcut from the toolbar. The SDK path that was configured in Android Studio is displayed on the top of the manager.

The SDK Manager displays the list of available packages with the following properties:

- **Name**: This is the name of the package or the container that aggregates related packages.
- **API**: This is the API number in which the package was added.
- **Rev.**: This is the package revision or version.
- **Status**: This is the status of the package on your system. The status can be **Not installed**, **Installed**, **Update available**, **Not compatible**, or **Obsolete**.

The packages can be filtered by their state using the checkboxes under the list, and they can be sorted by the API level or by the repository they are downloaded in. These options are also accessible from the **Packages** menu at the top.

By navigating to **Tools** | **Manage Add-on Sites**, we can examine the list of official sites that provide the add-ons and extra packages. We can add our custom external sites to the **User Defined Sites** tab.

Next to the name of the packages, there is a checkbox to select the packages we want to install, update, or delete. As shown in the next screenshot, the packages that are installed in our system and also have an update available are checked by default:

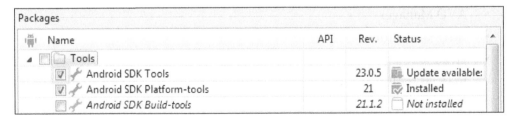

If there is a new Android platform version that is not installed, its packages will also be checked, as shown in the following screenshot:

Packages				
Name		API	Rev.	Status
▲ ☐ Android 5.0 (API 21)				
☑ Documentation for Android SDK		21	1	☐ Not installed
☑ SDK Platform		21	1	☐ Not installed
☑ Samples for SDK		21	4	☐ Not installed
☑ Android TV ARM EABI v7a System Image		21	1	☐ Not installed
☑ Android TV Intel x86 Atom System Image		21	1	☐ Not installed

The total number of selected packages to be installed or updated is indicated in the text of the button at the bottom of the dialog. The button under it indicates the total number of selected packages to be deleted. You can delete packages that are deprecated or packages that you do not need anymore.

Check the packages that need to be updated, and also check the last Android platform. In addition, you should check the minimum platform supported by our application (Android 4.0.3, API15) to be able to test our application in a virtual device using this version. Click on the **Install** button.

In the next dialog, we have to accept the package licenses. Check the **Accept License** radio button and click on the **Install** button. The installation or update of the packages will start showing its progress. First, the manager downloads the packages, then it unzips them, and finally, it installs them.

Remember to check out the SDK Manager from time to time for updates.

The AVD Manager

The AVD Manager is an Android tool accessible from Android Studio to manage the Android virtual devices that will be executed in the Android emulator.

To open the AVD Manager from Android Studio, navigate to **Tools | Android | AVD Manager**. You can also click on the shortcut from the toolbar. The AVD Manager displays the list of existing virtual devices. Since we have not created any virtual device, the list will initially be empty. To create our first virtual device, click on the **Create Virtual Device** button to open the configuration dialog.

Tools

The first step is to select the hardware configuration of the virtual device. The hardware definitions are listed in the left-hand side of the window. Select one of them, such as **Nexus 5**, to examine its details on the right-hand side, as shown in the following screenshot. Hardware definitions can be classified into one of these categories: **Phone**, **Tablet**, **Wear**, or **TV**.

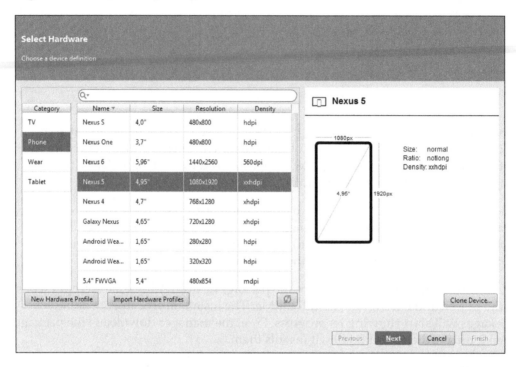

We can also configure our own hardware device definitions from the AVD Manager. We can create a new definition using the **New Hardware Profile** button. The **Clone Device...** button creates a duplicate of an existing device.

Click on the **New Hardware Profile** button to examine the existing configuration parameters. The most important parameters that define a device are:

- **Device Name**: This is the name of the device.
- **Screensize**: This is the screen size in inches. This value determines the size category of the device. Type a value of 4.0 and notice how the **Size** value (on the right-hand side) becomes **normal**. Now type a value of 7.0 and the **Size** field changes its value to **large**. This parameter, along with the screen resolution, also determines the **Density** category.

[58]

- **Resolution**: This is the screen resolution in pixels. This value determines the density category of the device. For a screen size of 4.0 inches, type a value of 768 x 1280 and notice how the **Density** value becomes **400 dpi**. Change the screen size to 6.0 inches and the **Density** value changes to **hdpi**. Now change the resolution to 480 x 800 and the **Density** value will be **mdpi**.
- **RAM**: This is the RAM memory size of the device.
- **Input**: This indicates whether the **home**, **back**, or **menu** buttons of the device are available via software or hardware.
- **Supported device states**: This checks the allowed states.
- **Cameras**: This checks whether the device has a front camera or a back camera.
- **Sensors**: These are the sensors available in the device. They are of the following types: **accelerometer**, **gyroscope**, **GPS**, and **proximity sensor**.
- **Default Skin**: This selects additional hardware controls.

Create a new device with a screen size of 4.7 inches, a resolution of 800 x 1280, a RAM value of 500 MiB, software buttons, and both portrait and landscape states enabled. Name it My Device. Then click on the **Finish** button. The hardware definition has been added to the list of configurations.

Click on the **Next** button to continue the creation of a new virtual device. The next step is to select the virtual device system image and the target Android platform. Each platform has its own architecture, so the system images that are installed on your system will be listed along with the rest of the images that can be downloaded (the **Show downloadable system images** box checked). Download and select one of the images of the Lollipop release, and click on the **Next** button.

The last step is to verify the configuration of the virtual device. Enter the name of the AVD in the **AVD Name** field. Give the virtual device a meaningful name to recognize it easily, such as AVD_nexus5_api21. Click on the **Show Advanced Settings** button. The settings that we can configure for the virtual device are the following:

- **Emulation Options**: The **Store a snapshot for faster startup** option saves the state of the emulator in order to load data faster the next time. The **Use Host GPU** option tries to accelerate the GPU hardware to run the emulator faster.
- **Custom skin definition**: Select this if additional hardware controls are displayed in the emulator.

Tools

- **Memory and Storage**: Select the memory parameters of the virtual device. Leave the default values as they are, but if a warning message is shown, follow the instructions of that message. For example, select **1536M** for the **RAM** memory and **64** for the **VM Heap** field. The **Internal Storage** option can also be configured, for example, **200 MiB**. Select the size of **SD Card**, or select a file to behave as the SD card.

- **Device**: Select one of the available device configurations. These configurations are what we tested in the layout editor preview. Select the **Nexus 5** device to load its parameters in the dialog.

- **Target**: Select the device Android platform. We have to create one virtual device with the minimum platform supported by our application, and another virtual device with the target platform of our application. For the first virtual device, select **Android 4.4.2 API19** as the target platform.

- **CPU/ABI**: Select the device architecture. The value of this field is set when we select the target platform. Each platform has its own architecture, so if we don't have it installed, the following message will be shown: **No system images installed for this target**. To solve this, open the SDK Manager and search for one of the architectures of the target platform, which could be **ARM EABI v7a System Image** or **Intel x86 Atom System Image**.

- **Keyboard**: Select this if a hardware keyboard is displayed in the emulator. Check it.

- **Skin**: Select this if additional hardware controls are displayed in the emulator. You can select the **Skin with dynamic hardware controls** option.

- **Front Camera**: Select this if the emulator has a front camera or a back camera. The camera can be emulated or can be real (by the use of a webcam from the computer). Select **None** for both cameras.

- **Keyboard**: Select this if a hardware keyboard is displayed in the emulator. Check it.

- **Network**: Select the speed of the simulated network and the delay in processing data across the network.

Chapter 6

The new virtual device is now listed in the AVD Manager. Select the recently created virtual device to enable the remaining actions:

- **Start**: This runs the virtual device.
- **Edit**: This edits the virtual device configuration.
- **Duplicate**: This creates a new device configuration displaying the last step of the creation process. You can change its configuration parameters and then verify the new device.
- **Wipe Data**: This removes the user files from the virtual device.
- **Show on Disk**: This opens the virtual device directory in your system.
- **View Details**: This opens a dialog detailing the virtual device's characteristics.
- **Delete**: This deletes the virtual device.

Click on the **Start** button. The emulator will be opened, as shown in the next screenshot. Wait until it is completely loaded, and then you will be able to try it.

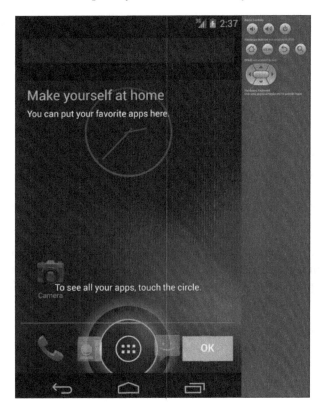

Tools

In Android Studio, open the main layout with the graphical editor and click on the list of devices. As the following screenshot shows, our custom device definition appears, and we can select it to preview the layout:

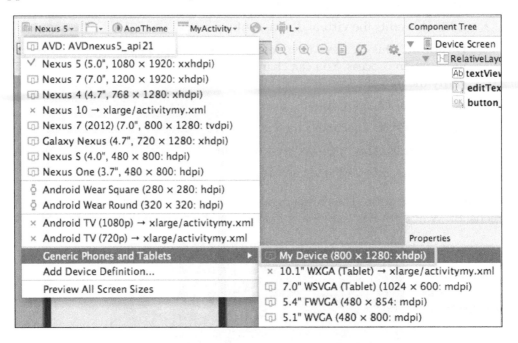

The Navigation Editor

The Navigation Editor is a tool used to create and structure the layouts of the application using a graphical viewer. To open this tool, navigate to **Tools** | **Android** | **Navigation Editor**. This tool opens a file in XML format, named `main.nvg.xml`. This file is stored in your project at `/.navigation/app/raw/`.

Since there is only one layout (and one activity) in our project, the navigation editor only shows this main layout. If you select the layout, detailed information about it is displayed on the panel on the right-hand side of the editor. If you double-click on the layout, the XML layout file will be opened in a new tab.

We can create a new activity by right-clicking on the editor and selecting the **New Activity** option. We can also add transitions from the controls of a layout by shift clicking on a control and then dragging to the target activity.

Open the main layout and create a new button with the `Open Activity` label:

```
<Button
        android:id="@+id/button_open"
        android:layout_width="wrap_content"
        android:layout_height="wrap_content"
        android:layout_below="@+id/button_accept"
        android:layout_centerHorizontal="true"
        android:text="Open Activity" />
```

Open the **Navigation** Editor and add a second activity. Now the Navigation Editor displays both activities, as shown in this screenshot:

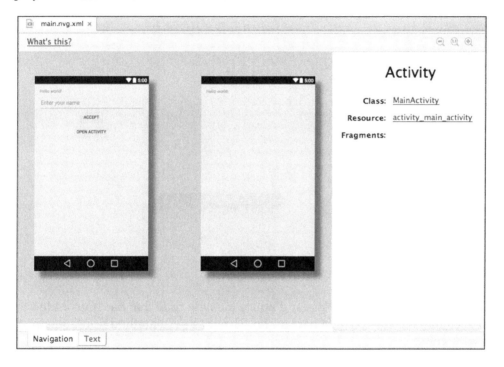

Tools

Now we can add the navigation between them. Shift-drag from the new button of the main activity to the second activity. A blue line and a pink circle have been added to represent the new navigation. Select the navigation relationship to see its details on the right panel, as shown in the following screenshot. The right panel shows the source of the activity, the destination activity, and the gesture that triggers the navigation.

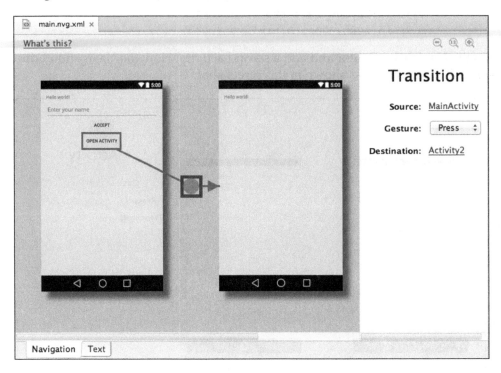

Now open our main activity class and notice the new code that has been added to implement the recently created navigation. The `onCreate` method now contains the following code:

```
findViewById(R.id.button_open).setOnClickListener(
new View.OnClickListener() {
   @Override
   public void onClick(View v) {
        MainActivity.this.startActivity(
           new Intent(MainActivity.this, Activity2.class));
   }
});
```

This code sets the `onClick` method of the new button, from which the second activity is launched.

Generating a Javadoc

A Javadoc is a utility to document Java code in HTML format. The Javadoc documentation is generated from comments and tags added to Java classes or methods. The comments start with the /** string and end with */. Inside these comments, tags can be added, such as @param to describe a method parameter, @throws to describe an exception that can be thrown by the method, or @version to indicate the version of the class or method.

The use of a Javadoc is integrated in Android Studio. We can use code completion when typing the Javadoc comments and the documentation will appear in the pop-up tool tips of the code elements.

To generate a complete Javadoc, we have to write the Javadoc comments about our classes and methods. Open the main activity of our project to add the Javadoc comments to the onAcceptClick method we created in *Chapter 5*, *Creating User Interfaces*. Place the caret on the line before the method declaration, type /**, and press *Enter*. The Javadoc comments are automatically inserted containing the available information from the method declaration: parameters and return type. In this case, there is no return type.

The first line of the documentation comments is the method description. Then, it explains each parameter and the return type. The method should now look like this:

```
/**
 * Method executed when the user clicks on the Accept button.
 * Change the greeting message to include the name introduced by the user in the editText box.
 *
 * @param v View the user clicked
 */
public void onAcceptClick(View v) { ... }
```

This information about the method will now be displayed as its documentation in the emerging dialogs. The following screenshot shows the dialog that should appear over the method:

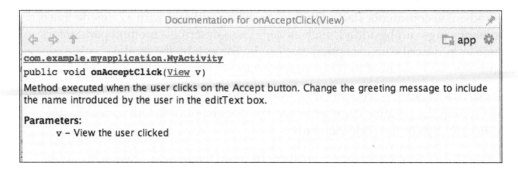

To generate the Javadoc documentation, navigate to **Tools** | **Generate Javadoc...**. A dialog showing the Javadoc options will be opened. We can choose the scope, output directory, and visibility of the included elements, and create a hierarchy tree, a navigation bar, and an index if needed.

Check **Current File** as the scope to generate only the documentation of our main activity. Select an output directory from your system. Reduce the visibility to **public** and click on the **OK** button. The Javadoc documentation in HTML format has been created in the output directory. The index.html file is the start point. Navigate through the documentation to open the MyActivity class. Notice that the onCreate method, whose visibility is protected, does not appear, as we reduced the visibility of the generated Javadoc to **public** elements.

Version control systems

Android Studio integrates some **version control systems** (**VCS**): GitHub, CVS, Git, Mercurial, and Subversion. To enable version control integration, navigate to **VCS** | **Enable Version Control Integration** and select the type of system. Now some more options will be added to the **VCS** menu:

- To update the entire project, navigate to **VCS** | **Update Project**
- To commit all the changes of the project, navigate to **VCS** | **Commit Changes**
- To clean up the project, navigate to **VCS** | **Cleanup Project**

The first step is to do the checkout from the version control system. Go to **VCS | Checkout from Version Control**, click on the add icon, and type the repository URL or repository configuration.

The version control actions can also be applied to individual files. Right-click on any file of the project and select the **Subversion** section. From the emerging menu, we can add the file to the repository, add it to the ignore list, browse the changes, revert the changes, or lock it.

A simpler way to control the file versions is by using the **Local History** option. Open the main activity file in the editor and navigate to **VCS | Local History | Show History**. The file history dialog will be opened. On the left-hand side of the dialog, the available versions of the file are listed. Select an older version to compare it to the current version of the file. The differences between the older version and the current version are highlighted. A gray color is used to indicate a block of deleted code, a blue color to highlight the text that has changed, and a green color to indicate the new inserted text. From the top icons, we can revert the changes and configure the white space visualization. The following screenshot shows the comparison between two versions of our main activity. We can observe how the method we recently added—the onAcceptClick method—is highlighted in green.

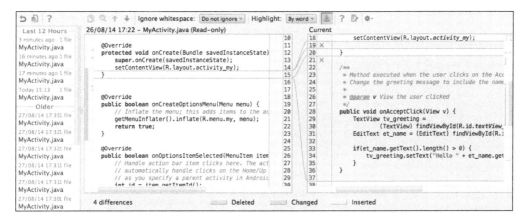

We can also examine the local history of a specific block of code. Close the dialog, select some lines of code from the editor, and go to **VCS | Local History | Show History for Selection**. The same history dialog will be opened, but this time, it displays the versions of the selected code.

Summary

By the end of this chapter, you have the knowledge required to use the Android SDK Manager tool to install, update, or examine available platforms for your project. You can create a new AVD and edit it whenever necessary. You learned how to use Navigation Editor, the new Android Studio tool. Creating complete documentation of our project should no longer be a problem using Javadoc, and we should also be able to work with a VCS integrated in Android Studio.

In the next chapter, we will keep on working with Android Studio integrated features. You will be learning about the emulation of your project and how to debug it. You will also learn about the debugger, the console, and the Logcat tool. Then you will learn about more advanced debugging tools such as the **Dalvik Debug Monitor Server (DDMS)**. We will study this monitor server in depth and go through each of its available utilities.

7
Google Play Services

Now that we have become familiar with the use of components on layouts, it is time to start thinking about extra functionality. Google Play services give you features such as Google Maps, Google+, and more, to attract users. What are all the features available? How can you add these features to your application? What are the Android version requirements to use Google Play services?

This chapter focuses on the creation, integration, and use of Google Play services using Android Studio. You will learn what Google services are available. You will also learn about the standard authorization API that provides a safe way to grant and receive access tokens from Google Play services. Then you will learn about the limitations of these services and also the benefits of using them.

These are the topics we'll be covering in this chapter:

- Existing Google services
- Adding Google Play services from the IDE
- Integrating Google Play services in your app
- Understanding automatic updates
- Using Google services in your app

How Google Play services work

When Google previewed Google Play services at Google I/O 2012, it said that the platform *"consists of a services component that runs on the device and a thin client library that you package with your app"* (https://developers.google.com/events/io/2012/).

This means that Google Play services work, thanks to two main components:

- **Google Play Client library**: The Google Play services client library includes interfaces to each Google service used by your app. The library is included when you pack your app, and it allows your users to authorize the app with access to these services using their credentials. The client library is upgraded from time to time by Google by adding new features and services. You may upgrade the library in your app through an update to your app, although this is not necessary if you are not including any of the new features.
- **Google Play services APK**: The Google Play services Android Package runs as a background service in the Android operating system. Using the client library, your app accesses this service, which carries out the actions during runtime. The APK is not guaranteed to be installed on all devices. If the device does not come installed with the APK, you can get it from the Google Play store.

In this way, Google manages to separate the runtime of their services from the implementation you do as a developer, so you do not need to upgrade your application every time Google Play services are upgraded.

Although Google Play services are not included in the Android platform itself, they are supported by most Android-based devices. Any Android device running Android 2.2 or newer is ready to install any application that uses Google Play services.

Services available

Google Play services let you easily add more features to attract users on a wide range of devices, while using well-known features powered by Google. Using these services, you can add new revenue sources, manage the distribution of the app, access statistics, learn about your application's users' habits, and improve your application with easy-to-implement Google features such as Maps or Google's social network, Google+. The services are explained as follows:

- **Games**: Using the Google Play Game service, you can improve your gaming with a more social experience.
- **Location**: By integrating the location APIs, you can make your application location-aware.
- **Maps**: By integrating the Google Maps API, you can use the maps provided by Google in your application and customize them.
- **Google+**: Using Google+ Platform for Android, you can authenticate the user of your app. Once they are authenticated, you can also access their public profile and social graph.

- **In-app Billing**: Using Google Play In-app Billing makes it possible for you to sell digital content from your apps. You can use this service to sell one-time billing or temporal subscriptions to premium services and features.
- **Cloud Messaging**: Using **Google Cloud Messaging (GCM)** for Android, you can exchange data between the app running in an Android-based device and your server.
- **Panorama**: By integrating this service, you can enable the user to see a 360-degree panoramic picture.
- **Analytics**: By integrating this service, you can allow your application to send information to Google Analytics.
- **Drive**: Using the Google Drive API, you can enable your application to access your users' files stored in their Google Drive accounts.
- **Wallet**: By integrating Google Wallet, you can store objects such as gift cards or loyalty programs in the cloud, and use them to pay in stores or online.

Adding Google Play services to Android Studio

The first thing we need to know is what we need to add to our Android Studio. We have just learned that the APK is available in the Google Play store and it is the actual runtime of the services. We, as developers, only need this package in our testing device while debugging our application. What we need to add to Android Studio is the Google Play services client library.

This library has to be declared as a dependency to your application, so perform the following steps:

1. Open the `build.gradle` file for your application module (`/app/build.gradle`).
2. Type the following line inside the `dependencies` block:

   ```
   dependencies {
       ...
       compile 'com.google.android.gms:play-services:5.+'
   }
   ```

   The latest Google Play services version is 5.0 (July 2014). When new updates of the library are published, you will need to update the version number in the `build.gradle` file.

3. Navigate to **Tools | Android | Sync Project with Gradle Files** to synchronize your project with the new dependency on Google Play services.

4. Finally, add it to the manifest file of your application under the `application` block:

   ```
   <meta-data android:name="com.google.android.gms.version"
       android:value="@integer/google_play_services_version" />
   ```

   You should have the library inside the `build` folder of your application project, as shown in the following screenshot:

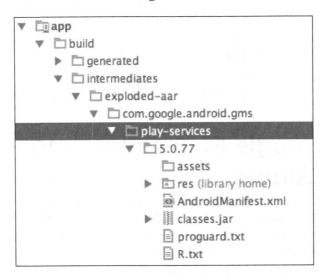

Another option is to download and install Google Play services in your system. This library is distributed through the **Android SDK Manager**, which will be explained in detail in *Chapter 7, Google Play Services*. Now perform the following steps:

1. Navigate to **Tools | Android | SDK Manager**. We can find Google Play services in the packages list under the **Extras** folder.

2. Select the **Google Play services** checkbox and click on the **Install 1 package...** button:

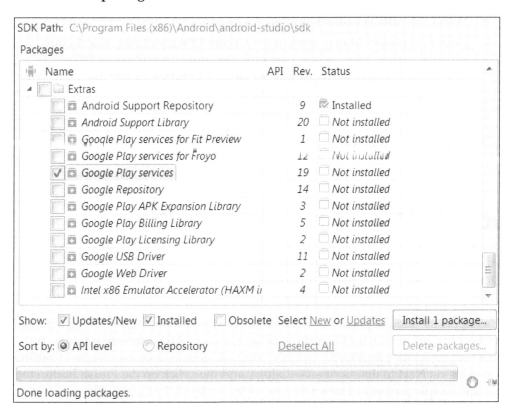

Performing these actions will add the library project to the location of our SDK installation folder, /sdk/extras/google/google_play_services/. You can check the exact path by hovering the mouse over the **Google Play services** row in the SDK manager and looking at the tool tip.

3. Navigate to the folder to examine its content. The samples folder contains some sample projects, for example, projects of the Analytics (analytics/), the authentication service (auth/), the Google Maps v2 service (maps/), the Panorama service (panorama/), the Google+ service (plus/), and Google Wallet (wallet/). The libproject/ folder contains the Google Play services library project. The google-play-services.jar file is placed in this folder at libproject/google-play-services_lib/libs/google-play-services.jar.

Google Play Services

4. Add this JAR file to your project by dragging it into the `libs/` folder.
5. Select the JAR file and right-click on it.
6. Select the **Add as Library** option.
7. Select the project library level in the **Create Library** dialog, select your application module, and click on **OK**. You now have the `google-play-services.jar` file available in your project libraries, under the `libs/` folder, and you will now be able to reference Google Play services from your code.
8. Finally, you will need to add the library to your Gradle's build file. To do this, just edit the `build.gradle` file under `MyApplication/`, and add the following line in the `dependencies` section:

   `compile files('libs/google-play-services.jar')`

Google Maps Android API v2

Google Maps Android API allows users of your application to explore maps available through a Google service. The new Maps version 2 offers more functionalities such as 3D maps, indoor and satellite maps, efficient caching and drawing using vector-based technology, and animated transitions through the map.

Let's import the sample project to examine its most important classes. Navigate to **File | Import Project**, search for the sample project in your SDK installation folder, and select the project root directory, `/google_play_services/samples/maps/`. Continue by clicking on **Next** in the successive dialogs, and then click on the **Finish** button to open the sample project in a new window. Now we have the Google Play services project and the `maps` sample project loaded in a new window in Android Studio.

Open the `BasicMapDemoActivity` class under the `maps` project under `src/main/java/`. The `com.google.android.gms.maps` package contains the Google Maps Android API classes. The `GoogleMap` class is the main class of the API, and it is the entry point for all the methods related to a map. You may change the theme colors and the icons of your map to match your application style. You can also customize your map by adding markers to it. To add a simple marker, you can use the `addMarker` method of the `GoogleMap` class. Examine the `onMapReady` method in `BasicMapDemoActivity` to see the following code:

   `mMap.addMarker(new MarkerOptions().position(new LatLng(0, 0)).title("Marker"));`

The `addMarker` method has a `MarkerOptions` object as a parameter. Using the `position` method, we indicate the coordinates of the marker on the map, and using the `title` method, we can add a custom string to show up on the marker.

To add a map to a layout, we can use the `MapView` class, which extends the `View` class and displays a map. However, the easiest way to place a map in an application is by using a `MapFragment` object. A **fragment** represents a piece of the user interface or behavior that can be embedded in an activity. A fragment is a reusable module.

The `MapFragment` class wraps a view of a map to handle the necessary life cycle requirements of a component automatically. It extends the `Fragment` class, so it can be added to a layout using the following XML code:

```
<fragment
    class="com.google.android.gms.maps.MapFragment"
    android:layout_width="match_parent"
    android:layout_height="match_parent" />
```

To see an example of this code in use, open the layout associated with the `BasicMapDemoActivity` class. This is the `basic_demo.xml` file found under `/res/layout/`.

Finally, we need the code to obtain the `GoogleMap` object from the fragment. We can find the `Fragment` map using the `findFragmentById` method, and then we can get the map from the `Fragment` map using the `getMap` method:

```
SupportMapFragment mapFragment = ((SupportMapFragment)
getSupportFragmentManager().findFragmentById(R.Id.map);
```

You can see an example of this code in the `BasicMapDemoActivity` class in the `onCreate` method.

The last important class to cover is the `GoogleMapOptions` class. It defines the configuration for a map. You can also modify the initial state of a map by editing the layout XML code. Here are some interesting options that are available:

- `mapType`: This specifies the type of map. Its value can be `none`, `normal`, `hybrid`, `satellite`, or `terrain`.
- `uiCompass`: This defines whether compass controls are enabled or disabled.
- `uiZoomControls`: This defines whether zoom controls are enabled or disabled.
- `cameraTargetLat` and `cameraTargetLong`: This specifies the initial camera position.

Google+ Platform for Android

Using Google+ Platform for Android lets a developer authenticate users using the same credentials that they use on Google+. It also enables the use of their public profile and social graph to welcome users by their name, display their pictures, and connect with their friends.

Import the Google+ sample project to learn about the most important classes. The Google+ sample project can be found in the Google Play services installation folder, at `/google_play_services/samples/plus/`. The `com.google.android.gms.samples.plus` package contains Google+ Platform for Android classes. The following are the classes found in this package:

- `PlusClient` and `PlusClient.Builder`: `PlusClient` is the main class of the API. It is the entry point for Google+ integration. `PlusClient.Builder` is a builder used to configure the `PlusClient` object to communicate properly with the Google+ APIs.

- `PlusOneButton`: This class implements a **+1** button to recommend a URL on Google+. Add it to a layout using the following code:

    ```
    <com.google.android.gms.plus.PlusOneButton
        android:layout_width="wrap_content"
        android:layout_height="wrap_content"
        plus:size="standard" />
    ```

    The available sizes are `small`, `medium`, `tall`, and `standard`.

    An example of code showing this functionality can be found in the sample project in the `PlusOneActivity` class in the `src/` folder. Its associated layout can be found in the `plus_one_activity.xml` file at `res/layout/`.

- `PlusShare`: This includes the resources in posts shared on Google+. An example code of sharing resources can be found in the `ShareActivity` class in the `src/` folder and its associated layout, `share_activity.xml`, in the `res/layout/` folder.

A `PlusClient` object should be instantiated in the `onCreate` method of your activity class to call its asynchronous `connect` method, which will connect the client to Google+ services. When the app is built using a `PlusClient` instance, it should call the `disconnect` method, which terminates the connection, and should always be called from the `onStop` method of the activity.

Google Play In-App Billing v3

In-app Billing v3 allows you to sell virtual content from your apps. This virtual content could be paid content with a one-time billing or a time concession through subscriptions or fees. Using this service, you can allow users to pay for extra features and access premium content.

Any app published in the Google Play store can implement the In-app Billing API, since it only requires the same assets as publishing an app: a Google Play Developer Console account and a Google Wallet Merchant account.

Using Google Play Developer Console, you can define your products, including the type, identification code (SKU), price, description, and more. Once you have your products defined, you can access this content from this application. When the user wants to buy this content, the following purchase flow will occur between your In-app Billing application and Google Play App:

1. Your app calls `isBillingSupported()` to Google Play to check whether the In-app Billing version you are using is supported.
2. If the In-app Billing API version is supported, you may use `getPurchases()` to get a list of the SKUs of the purchased items. This list will be returned in a `Bundle` object.
3. You will probably want to inform your user of the in-app purchases available. To do this, your app may send a `getSkuDetails()` request, which will result in a list with the product's price, title, description, and more information about the item being offered.

Google Cloud Messaging

Google Cloud Messaging (GCM) for Android allows communication between your server and your application using asynchronous messages. You don't have to worry about handling low-level aspects of this communication, such as queuing and message construction. Using this service, you can easily implement a notification system for your application.

You have two options when using GCM:

- The server can inform your app when there is new data available to be fetched from the server, and then the application gets this data.
- The server can send the data directly in a message. The message payload can be up to 4 KB. This allows your application to access the data at once and act accordingly.

In order to send or receive messages, you will need to get a registration ID. This registration ID identifies the combination of the device and the application. To allow your app to use the GCM service, you need to add the following line to the manifest file of your project:

```
<uses-permission android:name="com.google.android.c2dm.permission.
RECEIVE"/>
```

The main class you will need to use is `GoogleCloudMessaging`. This class is available in the `com.google.android.gms.gcm` package.

Summary

In this chapter, we discussed the available Google Play services. You learned how to improve your application using Google Play services through its client library and Android package. You should have successfully installed the Google Play services client library in Android Studio using the SDK Manager, and should now be able to build applications using the library features. You also learned some tips about Google Maps v2, Google+ Platform for Android authentication, Google Play In-app Billing, and GCM.

In the next chapter, you will learn about some useful tools available in Android Studio. We will use the SDK Manager frequently to install different packages. You will also learn about the AVD Manager for different virtual devices to test your applications on. We will generate Javadoc documentation for our project using the Javadoc utility, and you will learn about the version control systems available in Android Studio.

8
Debugging

The debugging environment is one of the most important features of an IDE. Using a debugging tool allows you to easily optimize your application and improve its performance. Do you want to use a debug tool while programming in Android Studio? Android Studio includes the **Dalvik Debug Monitor Server (DDMS)** debugging tool.

In this chapter, you will start by learning about the **run** and **debug** options, and how to emulate your application in one of the Android virtual devices you learned to create in the previous chapter. You will learn about the **Debugger**, **Console**, and **LogCat** tabs in depth. You will also learn how to use breakpoints when using the debugger. We will end this chapter with information about each tab available in the advanced debugger tool included in Android Studio DDMS.

These are the topics we'll be covering in this chapter:

- Debugging
- LogCat
- Device Monitor tools (DDMS)

Running and debugging

Android applications can be run from Android Studio in a real device using a USB connection or in a virtual device using the emulator. Virtual devices make it possible to test our applications on different types of hardware and software configurations. In this chapter, we will use the emulator to run and debug our application because of its simplicity and flexibility.

Debugging

To run an application directly, navigate to **Run | Run 'app'**. You can also click on the Play icon from the toolbar. To debug an application, navigate to **Run | Debug 'app'** or click on the Bug icon from the toolbar. When we select the **Debug 'app'** option, a dialog to choose the device is opened. The first option is to choose a running device; the available connected devices are listed, real, or virtual. The second option is to launch a new instance of the emulator; the available virtual devices are listed. Check the **Launch emulator** option, select the virtual device created in *Chapter 6, Tools*, and then click on **OK**. The emulator will be launched. The next time we run or debug the application, the emulator will be running, so we will choose the first option (**Choose a running device**), as shown in the following screenshot:

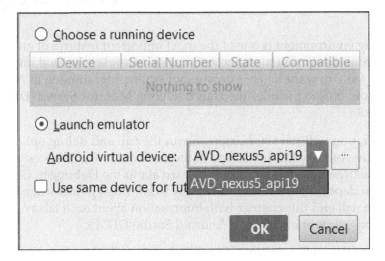

While debugging, you will notice that, at the bottom of Android Studio, there are three tabs contained in the **Debug** panel: **Debugger**, **Console**, and **LogCat**.

Console

Console displays the events that are taking place while the emulator is being launched. Open it to examine the messages and check that the emulator and the application are being correctly executed. The actions that should appear are:

- **Waiting for device**: This is the starting point when the emulator is being launched.
- **Uploading file**: This event states that the application is packed and stored in the device.
- **Installing**: This event states that the application is being installed in the device. After the installation, a success message should be printed.

- **Launching application**: This event takes place when the application starts to execute.
- **Waiting for process**: This event takes place when the application is running and the debug system is trying to connect to the application process in the device.

After the success of the previous steps, the application will be visible in the emulator. Test it by typing any name in the text input and clicking on the **Accept** button. The greeting message should change.

Debugger

Debugger manages the breakpoints, controls the execution of the code, and shows information about the variables. To add a breakpoint in our code, just click on the left edge of a line of code. A red point will appear next to the line of code to indicate the breakpoint. To delete the breakpoint, click on it. If you right-click on a breakpoint, more options become available. We can disable it without deleting it, or we can set a condition for the breakpoint.

Add a breakpoint in the conditional statement of the `onAcceptClick` method of our main activity and debug the application again, as shown:

```java
/**...*/
public void onAcceptClick(View v) {
    TextView tv_greeting = (TextView) findViewById(R.id.textView_greeting);
    EditText et_name = (EditText) findViewById(R.id.editText_name);

    if(et_name.getText().length() > 0) {
        tv_greeting.setText("Hello " + et_name.getText());
    }
}
```

Enter your name in the application and click on the **Accept** button. When the execution gets to the breakpoint, it pauses, and the **Debugger** tab is opened. Since we added the breakpoint in the conditional statement before assigning the text, our greeting message has not changed.

Debugging

From the **Debugger** tab, we can examine the method call hierarchy and the state of the variables at that point of execution. The available variables are the parameter of the v method, the `TextView` and `EditText` objects obtained by the `findViewById` method, and the reference to the current activity (`this`). Expand the `EditText` object named `et_name`, as shown in the following screenshot, and search for the `mText` property:

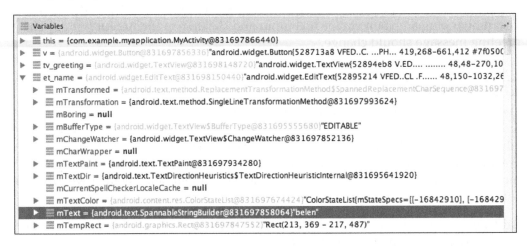

This property should contain the name you typed before:

- To execute the next line of code without stepping into the method call, navigate to **Run | Step Over** or use the keyboard shortcut indicated for this option, usually the *F8* key.
- To step into the method call, navigate to **Run | Step Into** or press *F7*.
- To resume the execution until the next breakpoint, provided there is any breakpoint, navigate to **Run | Resume Program** or press *F9*.
- To stop the execution, navigate to **Run | Stop** or press *Ctrl + F2*.

These options, among others, are also available from the **Debugger** tab as icon shortcuts.

Expand the `tv_greeting` object to check the value of its `mText` property. Now step over the conditional statement and the call of the `setText` method. Notice how the value of the `mText` property has changed, which is shown in the next screenshot. Finally, resume the execution so the greeting message changes in the device screen.

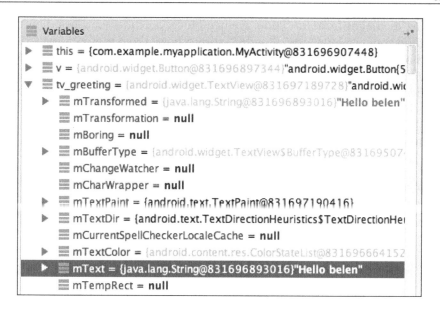

LogCat

LogCat is the Android logging system that displays all the log messages generated by the Android system in the running device. Log messages have several levels of significance. From the **LogCat** tab, we can filter log messages by these levels. For example, if we select the information level as the filter, the messages from `information`, `warning`, and `error` levels will be displayed. The levels are shown in the following diagram:

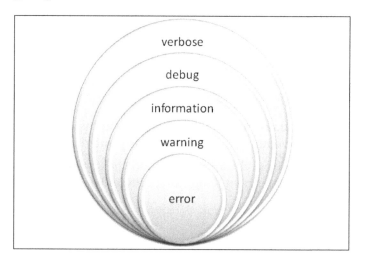

Debugging

To print log messages from our code, we need to import the Log class. This class has a method for each level: the v method for verbose, the d method for debug, the i method for information, the w method for warning, and the e method for the error level. These methods receive two string parameters. The first string parameter usually identifies the source class of the message, and the second string parameter identifies the message itself. To identify the source class, we recommend using a constant, static string tag. However, in the next example, we directly use the string to simplify the code. Add the following log messages to the onAcceptClick method of our main activity:

```
if(et_name.getText().length() > 0) {
  Log.i("MainActivity", "Name read: " + et_name.getText());
  tv_greeting.setText("Hello " + et_name.getText());
}
else {
    Log.w("MainActivity", "No name typed, greeting didn't change");
}
```

We have a log message to inform us about the name obtained from the user input, and a log message to print a warning if the user did not type any name. Remove any breakpoint we previously created and then debug the application.

The **LogCat** tab has printed all the log messages generated by the device, so reading the messages of our application can be complex. We need to filter the messages. In the **LogCat** tab, there is an expandable list with the **No Filters** option selected. Expand it and select the **Edit Filter Configuration** option. A dialog to create filters is opened. Log messages can be filtered by their tag or their content using regular expressions, by the name of the package that printed them, by the **process ID (PID)**, or by their level.

Create a new filter named MyApplication, filter it by **Package Name** writing com.example.myapplication (our application package name), and click on **OK**. Now the **LogCat** log has been filtered, and it is easier to read our messages. Now perform the following steps:

1. Focus on the **Emulator** window, enter a name in the application, and click on **Accept**. Observe how our log message is printed in the **LogCat** view.

2. Delete your name from the application and click on **Accept**. This time, a warning message is printed. Notice the different colors used for each type of message.

Memory Monitor

Memory Monitor is available at the bottom-right corner of Android Studio. You can also navigate to **Tools | Android | Memory Monitor**. Select the device or emulator running your application, and select the process corresponding to your application.

The **Memory Monitor** tab shows the free and allocated memory of the selected application over time, as shown in the following screenshot:

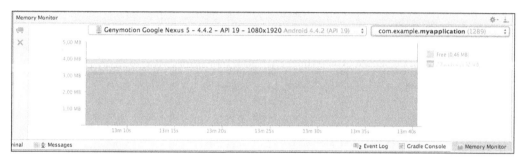

Android Device Monitor

The **Dalvik Debug Monitor Server (DDMS)** is a more advanced debugging tool available in the SDK. The DDMS can be accessed from Android Studio through the **Android Device Monitor** tool. This tool is able to monitor both a real device and the emulator.

To open the DDMS perspective, navigate to **Tools | Android | Android Device Monitor**. You can also click on the Android Device Monitor icon from the toolbar. A new window will be opened with the DDMS perspective.

In the left part of the window, the list of connected devices is shown. Currently, just our virtual device is listed. In the **Devices** section, the list of the processes running on each device is also presented. We should be able to locate our application in the processes of the device we launched before. From the toolbar of the **Devices** section, we can stop a process using the Stop sign icon. We can also take a screen capture of the virtual device by clicking on the Camera icon. Some of the other options will be explained later.

In the right part of the window, detailed information about the device is provided. This information is divided into seven tabs: **Threads**, **Heap**, **Allocation Tracker**, **Network Statistics**, **File Explorer**, **Emulator Control**, and **System Information**. **LogCat**, which has been also integrated in the DDMS perspective, is placed at the bottom of the window.

Threads

The **Threads** tab displays the list of threads that belong to the selected process. Select our application process from the **Devices** section. The process is identified by the package name, com.example.myapplication in this case. Click on the Update Threads icon button from the toolbar of the **Devices** section and the threads will be loaded in the content of the tab:

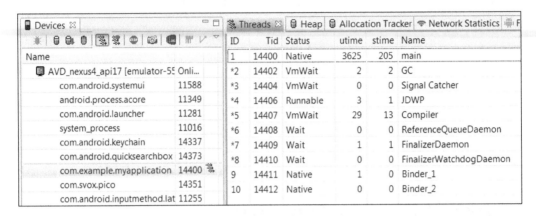

The first columns are the IDs of the threads. The **Status** column indicates the thread state, **utime** indicates the total time spent by the thread executing the user code, **stime** indicates the total time spent by the thread executing the system code, and **Name** indicates the name of the thread. The threads that interest us are those that spend time executing our user code.

This Threads tool is useful if we create threads in our application apart from the main thread. We can check whether they are being executed at a certain point of the application or whether their execution time is moderate or not.

Method profiling

Method profiling is a tool used to measure the performance of the method execution in the selected process. The measured parameters are the number of calls and the CPU time spent while executing. There are the following two types of spent time:

- **Exclusive time**: This is the time spent in the execution of a method.

- **Inclusive time**: This is the total time spent in the execution of a method. This measure includes the time spent by any called method inside the method. These called methods are known as its children methods.

To collect the method profiling data, select our application process from the **Devices** section, and click on the Start Method Profiling icon from the toolbar of the **Devices** section, next to the Update Threads icon. Then perform some actions in the application; for example, in our example application, type a name and click on the **Accept** button in order to execute the onAcceptClick method of the main activity. Stop the method profiling by clicking on the Stop Method Profiling icon.

When the method profiling is stopped, a new tab with the resultant trace is opened in the DDMS perspective. On the top of this new tab, the method calls are represented in a time graph; each row belongs to a thread. On the bottom of the trace, the summary of the time spent in a method is represented in a table.

Order the methods by their name to search for our onAcceptClick method. Click on it to expand the detailed information about its execution. Now notice the following facts:

- The children methods called inside the onAcceptClick method are listed. We can see the EditText.getText method, the Activity.findViewById method, and the TextView.setText method, which we indeed directly call inside the method, as shown in the next screenshot.
- The number of calls is detailed. For example, we can see that the Activity.findViewById method is called twice: one call to find the TextView object, and a second call to find the EditText object.

Debugging

- The **Exclusive** time columns have no values for the parent or children methods due to their own definition of this type of measured time.

The following screenshot demonstrates the preceding points:

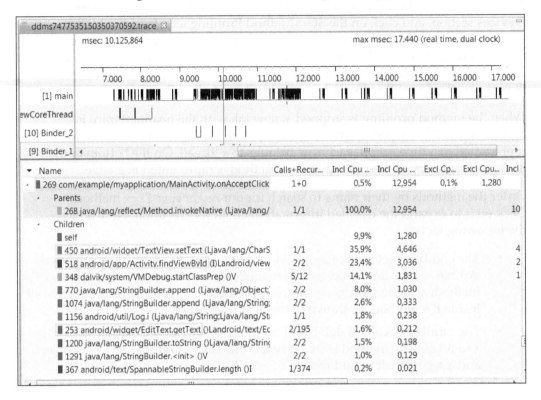

Method profiling is very useful to detect methods that spend too much time in their execution and to subsequently optimize them. We can identify the most expensive methods to avoid unnecessary calls to them.

Heap

The **Heap** tab displays the heap memory usage information and the statistics of the selected process. Select the application process and click on the Update Heap icon from the toolbar of the **Devices** section to enable it. The heap information is shown after a **garbage collector** (GC) execution. To force it, click on the **Cause GC** button or the Garbage icon from the toolbar of the **Devices** section.

The first table displays the summary of the heap usage: the total size, the allocated space, the free space, and the number of allocated objects. The **Stats** table gives the following details of the objects allocated in the heap by type: the number of objects, the total size of those objects, the size of the smallest and largest objects, the median size, and the average size. Select one of the types to load the bottom bar graph. The graph shows the count of the objects of a type by size, in bytes. If we right-click on the graph, we can change its properties (title, color, font, labels, and so on) and save it as an image in the PNG format.

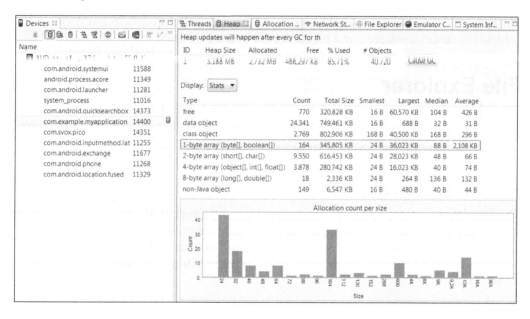

Allocation Tracker

The **Allocation Tracker** tab displays the memory allocations of the selected process. Select the application process and click on the **Start Tracking** button to start tracking the memory information. Then click on the **Get Allocations** button to get the list of allocated objects.

We can use the filter on the top of the tab to filter the objects allocated in our own classes. Type our package name com.example.myapplication in the filter. For each object, the table shows its allocation size, the thread, the object or class, and the method in which the object was allocated. Click on any object to see more information, for example, the line number that allocated it. Finally, click on the **Stop Tracking** button.

Debugging

The allocation tracker is very useful to examine the objects that are being allocated when doing certain interactions in our application, in order to improve memory usage.

Network Statistics

The **Network Statistics** tab displays how our application uses the network's resources. To get the network statistics of any application that uses the network, click on the **Start** button. The data transfers will begin to appear in the graph.

The network statistics are useful to optimize the network requests in our code and control the data transferred at a certain point of the execution.

File Explorer

The **File Explorer** tab exposes the whole filesystem of the device. We can examine its size, date, or the permissions of each element. Navigate to `/data/app/` to search for our `com.example.myapplication.apk` application package file.

Emulator Control

The **Emulator Control** tab allows us to emulate some special states or activities in the virtual device. We can test our application in different environments and situations to check whether it behaves as expected. If our application has features that depend on the device's physical location, we can use mock locations. Some of these special states are:

- **Telephony Status**: This allows you to choose the voice and data status and its speed and latency
- **Telephony Actions**: This is used to simulate an incoming call or SMS
- **Location Controls**: This is used to set the geolocation of the device

System Information

The **System Information** tab presents the frame render time, total CPU load, and total memory usage of the device as graphs. We can search for our application and easily compare it with the rest of the processes running on the device.

We can change the properties of the graphs such as colors, font, and title, and we can save them as images in the PNG format. To open these options, right-click on the graph elements.

Open the CPU load and save the graph while our application is running in the foreground. Then close the application and update the CPU load by clicking on the **Update from Device** button. Notice the difference between both graphs and notice the growth of the idle percentage, as shown in the following screenshot:

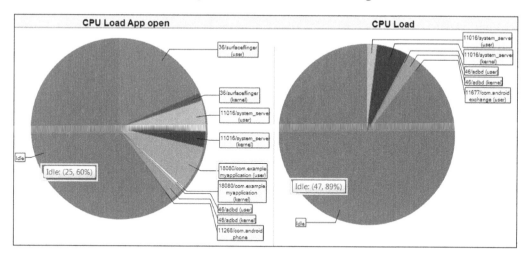

Summary

Now you know the different launch options for your application as well as how to use the console and the **LogCat** for debugging. We also saw how to debug an application and interpret the data provided by the DDMS in each of the tabs available.

In the next chapter, we will prepare our application for its release using Android Studio. First, you will learn the necessary steps to prepare the application before building it in release mode. You will also learn how the applications are compressed in APK files and how to generate your own APK file. Finally, you will learn how to get your certificate as a developer and how to generate a signed APK file, making it ready for release.

9
Preparing for Release

In the previous chapter, you learned enough to test and debug your application. What do you need to do to prepare your application for its release? How can you do this using Android Studio?

This chapter describes the necessary steps to prepare your application for release using Android Studio. First of all, you will learn about **application package** (**APK**) files—a variation of the JAR files in which Android applications are packed. You will then learn how you need to change your application after fully testing it. Finally, we will sign our APK file, leaving it ready to upload to Google Play.

These are the topics we'll be covering in this chapter:

- Preparing for release
- APK files
- Creating a certificate
- Generating a signed APK

Understanding an APK file

Android applications are packed in a file with the `.apk` extension. These files are just compressed ZIP files, so their content can be easily explored. An APK file usually contains the following:

- `assets/`: This is a folder that contains the asset files of the application. This is the same `assets` folder that exists in our project.
- `META-INF/`: This is a folder that contains our certificates.
- `lib/`: This is a folder that contains compiled code, in case it is necessary for a processor.

- `res/`: This is a folder that contains the application resources such as images, strings, and so on.
- `AndroidManifest.xml`: This is the application manifest file.
- `classes.dex`: This is a file that contains the application's compiled code.
- `resources.arsc`: This is a file that contains some precompiled resources such as binary XML files.

Having the APK file allows the application to be distributed and installed on the Android operating system. Android applications can be distributed as you prefer: through app markets such as Google Play, Amazon App Store, or Opera Mobile Store; through your own website; or even via an e-mail to your users. If you choose either of the two last options, take into account that Android, by default, blocks installations from locations different from Google Play. You should inform your users that they need to disable this restriction in their devices to be able to install your application. They have to check the **Unknown sources** option by navigating to **Settings | Security** in their Android devices.

Applications have to be signed with a private key when they are built. An application can't be installed in a device or even in the emulator if it is not signed. To build our application, there are two modes: **debug** and **release**. Both APK versions contain the same folders and compiled files; the difference is in the key used to sign them. Both modes are explained as follows:

- **Debug**: When we ran and tested our application in the previous chapters, we were in debug mode, but we didn't have a key nor did we do anything to sign our application. The Android SDK tools automatically create a debug key, an alias, and their passwords to sign the APK. This process occurs when we are running or debugging our application with Android Studio without us realizing it. We can't publish an APK signed with the debug key created by the SDK tools.
- **Release**: When we want to distribute our application, we have to build a release version. Google Play requires the APK file to be signed with a certificate, for which the developer keeps the private key. In this case, we need our own private key, alias, and password, and need to provide them to the build tools. The certificate identifies the developer of the application and can be a self-signed certificate. It is not necessary for a certificate authority to sign the certificate.

Keep the key store with your certificate in a secure place. To upgrade your application, you have to use the same key in order to upload the new version. If you lose the key store, you won't be able to update your application. You will have to create a new application with a different package name.

Steps to take before releasing your app

Before we generate the APK file, it is necessary to prepare our application to build it in release mode. Perform the following steps:

1. Firstly, make sure you have completely tested your application. We recommend testing your application in the following ways:
    - On a device using the minimum required platform
    - On a device using the target platform
    - On a device using the latest available platform
    - On a real device and not just the emulator
    - On a variety of screen resolutions and sizes
    - On a tablet if your application supports it
    - By switching to landscape mode if you can allow it, both in a mobile device and in a tablet
    - On different network conditions, such as with no Internet connectivity or low coverage
    - When the GPS or any location service is not activated on your device (if your application uses GPS or any location service)
    - When the Back button is pressed

2. Secondly, we have to check the log messages that are printed from our application. Printing some log messages can be considered a security vulnerability. Logs generated by the Android system can be captured and analyzed, so we should avoid showing critical information about the application's internal working. You should also remove the `android:debuggable` property from the application manifest file. You can also set this property to `false`.

3. Thirdly, if your application communicates with a server, check whether the configured URL is the production URL. It is possible that during the debug phase, you referenced to a URL of a server in a prerelease environment.

4. Finally, set the correct value for the `android:versionCode` and `android:versionName` properties from the application manifest file. The version code is a number (integer) that represents the application version. New versions should have greater version codes. This code is used to determine whether an application installed on a device is of the latest version, or if there is a newer version.

Preparing for Release

The version name is a string that represents the application version. Unlike the version code, the version name is visible to the user and appears in the public information about the application. It is just an informative version name to the user and is not used for any internal purpose.

Specify a value of 1 for the version code and 1.0 for the version name. The manifest tag should look like the following:

```
<manifest xmlns:android="http://schemas.android.com/apk/res/android"
    package="com.example.myapplication"
    android:versionCode="1"
    android:versionName="1.0" >
```

A new version of our application will have a value of 2 for the version code, and it could have 1.1 for the version name:

```
<manifest xmlns:android="http://schemas.android.com/apk/res/android"
    package="com.example.myapplication"
    android:versionCode="2"
    android:versionName="1.1" >
```

Generating a signed APK

To generate the signed APK, navigate to **Build | Generate Signed APK**. Select the app module and click on the **Next** button. In the dialog to generate the signed APK, we are asked for a certificate. The APK is signed by this certificate, which indicates that it belongs to us.

If this is our first application, we might not have any certificates. Click on the **Create new** button to open the dialog to create a new key store. Now fill in the following information:

- **Key store path**: This is the path in your system to create the key store. The key store is a file with the .jks extension, for example, release_keystore.jks.
- **Password**: This is the key store password. You have to confirm it.
- **Alias**: This is the alias for your certificate and is a pair of public and private keys. Let's name it releasekey.
- **Password**: This is the certificate password. You have to confirm it.

- **Validity (years)**: This is the certificate that will be valid until the validity date. A value of 25 years or more is recommended.
- **Certificate**: This is the personal information contained in the certificate. Type your first and last name, organizational unit, organization, city or locality, state or province, and country code; for example, AS as Organizational Unit, packtpub as Organization, and ES as Country Code.

Click on **OK**. The dialog to create the signed APK is now loaded with the key store data. The next time we create a signed APK, we will already have a certificate, so we will select the **Choose existing** button. Click on the **Next** button. In the next step, select the path to save the APK file, select the release build type, and click on **Finish**. When the APK is completely generated, we will be informed by a message on the bottom bar of Android Studio. We should have the APK file created in the selected path.

Now that you have the APK file ready for release, it is recommended that you test it again in a device before distributing it.

Summary

You learned how to make an APK file and how to modify your application to make it ready for release. You also learned how to sign your application using the developer certificate. By the end of this chapter, you should have generated a signed APK prepared for its release.

In *Appendix, Getting Help,* you will learn how to get help using Android Studio. We will access the Android Studio online documentation and go through the help topics. Finally, you will learn about keeping your Android Studio instance updated using the inbuilt feature for it.

Getting Help

When developing applications in a new IDE, there will always be doubts on how to perform a certain action. A successful IDE usually includes help wizards and documentation that help you with different problems. Have you wondered how to get help using Android Studio?

In this appendix, you will learn about Android Studio documentation and help topics. You will also learn about the topics available in the official documentation. They can be accessed online at the official Android website. Finally, you will learn how to keep your Android Studio instance up to date using the update functionality.

The following topics will be covered in this appendix:

- Android Studio help
- Online documentation
- Android Studio updates

Getting help from Android Studio

Android Studio documentation is included in the IntelliJ IDEA web help. This documentation is accessible from Android Studio by navigating to **Help** | **Online Documentation,** or at http://tools.android.com/welcome-to-android-studio. Another option is to navigate to **Help** | **Help Topics** to directly open the documentation contents tree, or go to http://www.jetbrains.com/idea/webhelp/intellij-idea.html. There are also some online video tutorials available. Navigate to **Help** | **JetBrains TV** or refer to http://tv.jetbrains.net/.

To quickly find actions of Android Studio, we can navigate to **Help** | **Find Action**. Type the action you are looking for, and the list of matching actions will be displayed.

Getting Help

Android Studio provides a tip of the day functionality. The tip of the day explains, in a dialog, a trick about Android Studio. Every time you open Android Studio, this dialog is shown. We can navigate through more tips using the **Previous Tip** and **Next Tip** buttons. By deselecting the **Show Tips on Startup** checkbox, we can disable this functionality. The tip dialog can be opened by navigating to **Help | Tip of the Day**.

Android online documentation

The official Android documentation provided by Google is available at `http://developer.android.com/`. This documentation contains all the necessary guides to learn not only how to program Android applications but also how to design for Android and distribute and promote our applications. Since this website is quite extensive, we are listing here some of the specific guides useful to increase the knowledge you will gain in the chapters of this book:

- *Chapter 1, Installing and Configuring Android Studio*:
    - The *Android Studio Overview* page at `http://developer.android.com/tools/studio/index.html`
    - The *Android Studio Tips and Tricks* page at `http://developer.android.com/sdk/installing/studio-tips.html`
    - The *Known issues* page at `http://tools.android.com/knownissues`

- *Chapter 2, Starting a Project*:
    - The *Launcher* section of the *Iconography* page at `http://developer.android.com/design/style/iconography.html#launcher`
    - The *Using Code Templates* page at `http://developer.android.com/tools/projects/templates.html`

- *Chapter 3, Navigating a Project*:
    - The *Managing Projects Overview* page at `http://developer.android.com/tools/projects/`

- *Chapter 4, Using the Code Editor*:
    - The *Keyboard Commands* section in the *Android Studio Tips and Tricks* at `http://developer.android.com/sdk/installing/studio-tips.html#KeyCommands`

- *Chapter 5, Creating User Interfaces*:
    - The *Layouts* page at http://developer.android.com/guide/topics/ui/declaring-layout.html
    - The *Input Controls* page at http://developer.android.com/guide/topics/ui/controls.html
    - The *Input Events* page at http://developer.android.com/guide/topics/ui/ui-events.html
    - The *Supporting Multiple Screens* page at http://developer.android.com/guide/practices/screens_support.html
- *Chapter 6, Tools*:
    - The *SDK Manager* page at http://developer.android.com/tools/help/sdk-manager.html
    - The *Managing Virtual Devices* page at http://developer.android.com/tools/devices/
- *Chapter 7, Google Play Services*:
    - The *Google Play Services* page at http://developer.android.com/google/play-services/
    - The *PlusOneButton* page at https://developer.android.com/reference/com/google/android/gms/plus/PlusOneButton.html
- *Chapter 8, Debugging*:
    - The *Using DDMS* page at http://developer.android.com/tools/debugging/ddms.html
    - The *Reading and Writing Logs* page at http://developer.android.com/tools/debugging/debugging-log.html
    - The *Profiling with Traceview and dmtracedump* page at http://developer.android.com/tools/debugging/debugging-tracing.html
- *Chapter 9, Preparing for Release*:
    - The *Publishing Overview* page at http://developer.android.com/tools/publishing/publishing_overview.html

Updates

From the **Help** menu, we can check for updates of Android Studio. Navigate to **Help | Check for Update**. If there is an update available for Android Studio that we have not installed, the update information will be shown in a dialog when the checking finishes. This dialog is shown in the next screenshot. We can see our current version, the new version code, and its size. We can choose to ignore the update, update it later (using the **Remind Me Later** button), review the online release notes about the update (using the **Release Notes** button), or install the update (using the **Update and Restart** button). Click on this last option to update Android Studio. The update starts to download first, then Android Studio will restart and the update will be installed.

> A new Android Studio 1.0.2 is available in the beta channel.
>
> Current version: 1.0.0 (build 135.1629389)
> New version: 1.0.2 (build 135.1653844)
> Patch size: 3 MB
>
> To configure automatic update settings, see the Updates dialog of your IDE Preferences.

If we already have the latest version of Android Studio, the following message will be shown:

> You already have the latest version of Android Studio installed.
>
> To configure automatic update settings, see the Updates dialog of your IDE Preferences.

Click on the **Updates** link to open the **Update Info** dialog. If we want, we can instruct Android Studio to automatically check for updates and the type of updates to check for, for example, beta releases or stable releases.

We can examine the information about the recent Android Studio updates by navigating to **Help | What's New** in Android Studio. This information is available online at http://tools.android.com/recent. To get the current version of Android Studio, or even the Java version in our system, navigate to **Help | About**.

Summary

You learned how to use the Android Studio documentation in case you need help with any action available in the IDE. You also learned about the update feature with which you can always install the latest version of Android Studio. By the end of this appendix, you should be able to search for help using the online documentation and the help topics, and keep your Android Studio updated with the latest features at your disposal.

Index

A

action bars
  URL 16
activities
  about 16
  Blank Activity 16
  Blank Activity with Fragment 16
  Fullscreen Activity 17
  Google Maps Activity 17
  Google Play Services Activity 18
  Login activity 18
  Master/Detail Flow 19
  Navigation Drawer Activity 19
  Settings Activity 20
  Tabbed Activity 20
allocation tracker 89, 90
Android
  dashboards, URL 15, 48
  developer tools, web page 6
  devices 15
  Google+ Platform for 76
  multiple screens, supporting 47-49
  screen sizes 49
Android Device Monitor tool
  about 85
  Allocation Tracker tab 89, 90
  Emulator Control tab 90
  File Explorer tab 90
  Heap tab 88, 89
  method profiling 86-88
  Network Statistics tab 90
  System Information tab 90, 91
  threads tab 86

Android online documentation
  URL 100, 101
Android Studio
  configuring 8-10
  documentation 99
  downloading 6
  Google Play services, adding 71-74
  installation, preparing for 5
  installing 6
  Quick Start section 7
  running 7
Android Virtual Device. *See* **AVD Manager**
APK file
  about 93
  AndroidManifest.xml file 94
  assets/ folder 93
  classes.dex file 94
  lib/ folder 93
  META-INF/ folder 93
  res/ folder 94
  resources.arsc file 94
application
  about 94
  debug 94
  releasing 94
  releasing, prerequisites 95, 96
application package. *See* **APK file**
application programming interface (API) 15
AVD Manager
  about 55-62
  New Hardware Profile button 58, 59
  Show Advanced Settings button 59, 60

B

Blank Activity
  with Fragment 16

C

code
  generating 37
  navigating 37-39
code completion
  about 34-36
  completion of statements 36
components
  adding 45, 46
Components Palette 43
Components Tree 43
component tree view and layout preview, differences
  hint 45
  id 45
  layout:width 45
  text 45
console
  about 80
  installing 80
  Launching application 81
  Uploading file 80
  Waiting for device 80
  Waiting for process 81
Containers, Components Palette 43
Custom, Components Palette 43
custom region
  about 38
  creating 38

D

Dalvik Debug Monitor Server (DDMS) 79, 85
Date & Time, Components Palette 43
debugger 81, 82
debugging 79, 80
density-independent pixel (dp) 49
device orientation 48

devices, Android
  Glass 15
  Phone and Tablet 15
  TV 15
  Wear 15
domain-specific language (DSL) 28

E

editor settings
  Appearance 33
  Auto Import 34
  Change font size (Zoom) with Ctrl+Mouse Wheel 32
  Code Completion 34
  Code Folding 33
  Colors & Fonts 33
  customizing 32, 33
  Editor Tabs 33
  Show line numbers, Appearance 33
  Show method separators, Appearance 33
  Show quick doc on mouse move 32
  Smart Keys 33
Emulator Control tab
  about 90
  Location Controls 90
  Telephony Actions 90
  Telephony Status 90
events
  handling 51-53
  OnClickListener 51
  OnCreateContextMenu 51
  OnDragListener 51
  OnFocusChange 51
  OnKeyListener 52
  OnLongClickListener 52
  OnTouchListener 52
Expert, Components Palette 43
extra-extra-high-density dots per inch (xxhdpi) 48
extra-high-density dots per inch (xhdpi) 48

F

File Explorer tab 90
form factors
  selecting 15

fragment
  about 75
  URL 16
Fullscreen Activity 17

G

Google Cloud Messaging (GCM) 77, 78
Google Maps Activity 17
Google Maps Android API v2 74, 75
Google+ Platform
  for Android 76
Google Play
  In-App Billing v3 77
Google Play services
  Activity 18
  adding, to Android Studio 71-74
  Analytics 71
  APK 70
  available 70, 71
  client library 70
  Cloud Messaging 71
  Drive 71
  Games 70
  Google+ 70
  In-app Billing 71
  Location 70
  Maps 70
  Panorama 71
  Wallet 71
  working 69, 70
Gradle
  about 28
  dependencies 28
  manifest entries 28
  signing 28
  URL 28
  variants 28
graphical editor 42

H

Heap tab 88, 89
high-density dots per inch (hdpi) 48
Holo style 50

J

Java Development Kit (JDK) 5
Javadoc
  generating 65, 66

L

layout
  new layout, creating 43, 44
LogCat 83, 84
Login activity 18
low-density dots per inch (ldpi) 48

M

Master/Detail Flow 19
medium-density dots per inch (mdpi) 48
medium high density dots per inch
    (tvdpi) 48
Memory Monitor 85
method profiling 86-88
module
  URL 29
multiple screens
  supporting 47-49

N

Navigate menu, custom region
  Call Hierarchy 39
  Class/File/Symbol 38
  File Path 39
  File Structure 39
  Last Edit Location 38
  Line 38
  Method Hierarchy 39
  Next Highlighted Error 39
  Next Method 39
  Previous Highlighted Error 39
  Previous Method 39
  Test 39
  Type Hierarchy 39
Navigation Drawer Activity
  about 19
  URL 19

[107]

Navigation Editor tool 62-64
navigation panel, project 24, 25
Network Statistics tab 90

O

OnClickListener event 51
OnCreateContextMenu event 51
OnDragListener event 51
OnFocusChange event 51
OnKeyListener event 52
OnLongClickListener event 52
OnTouchListener event 52

P

project
  activity type, selecting 16-20
  Application name 14
  Company Domain 14
  configuring 14
  creating 14
  form factors, selecting 15
  navigation panel 24, 25
  Package name 15
  Project location 15
  structure 26, 27
project, settings
  about 29
  Code Style 29
  Compiler 29
  File Encodings 29
  Gradle 29
  Language Injections 29
  Version Control 29
project, structure
  about 26
  AndroidManifest.xml file, src/main/ folder 27
  build/ folder 26
  build.gradle file 27
  Gradle 28
  java/ folder, src/main/ folder 26
  Libraries 30
  libs/ folder 26
  Modules 29
  Project 29
  res/ folder, src/main/ folder 27
  resources folder 27
  SDK Location 29
  src/androidTest/ folder 26
  src/main/ folder 26
Properties inspector 43

Q

Quick Start section 7, 8

R

relative layout 43
resources folder
  color/ 27
  drawable/ 27
  layout/ 27
  menu/ 27
  values/ 27

S

SDK Manager
  about 56, 57
  API 56
  Name 56
  Rev 56
  Status 56
Settings Activity 20
shortcuts
  about 40
  Alt + Arrows 40
  Ctrl + / 40
  Ctrl + A 40
  Ctrl + Alt + I 40
  Ctrl + Alt + O 40
  Ctrl + D 40
  Ctrl + F 40
  Ctrl + R 40
  Ctrl + Shift + U 40
  Ctrl + W 40
  Ctrl + Y 40
  Shift + Ctrl + Arrows 40
  Tab 40

signed APK
  alias 96
  certificate 97
  certificate password 96
  generating 96, 97
  key store password 96
  key store path 96
  validity (years) 97
smart type code completion 35, 36
software development kit (SDK) 55
System Information tab 90, 91

T

Tabbed Activity 20
text-based editor 44
Text Fields, Components Palette 43
threads tab 86
toolbar 44

U

UI theme
  changing 50
updates 102
user interface (UI) components
  about 43
  Containers 43
  Custom 43
  Date & Time 43
  Expert 43
  Layouts 43
  Text Fields 43
  Widgets 43

V

version control systems (VCS) 66, 67

W

Widgets, Components Palette 43

Thank you for buying
Android Studio Essentials

About Packt Publishing

Packt, pronounced 'packed', published its first book, *Mastering phpMyAdmin for Effective MySQL Management*, in April 2004, and subsequently continued to specialize in publishing highly focused books on specific technologies and solutions.

Our books and publications share the experiences of your fellow IT professionals in adapting and customizing today's systems, applications, and frameworks. Our solution-based books give you the knowledge and power to customize the software and technologies you're using to get the job done. Packt books are more specific and less general than the IT books you have seen in the past. Our unique business model allows us to bring you more focused information, giving you more of what you need to know, and less of what you don't.

Packt is a modern yet unique publishing company that focuses on producing quality, cutting-edge books for communities of developers, administrators, and newbies alike. For more information, please visit our website at www.packtpub.com.

About Packt Open Source

In 2010, Packt launched two new brands, Packt Open Source and Packt Enterprise, in order to continue its focus on specialization. This book is part of the Packt Open Source brand, home to books published on software built around open source licenses, and offering information to anybody from advanced developers to budding web designers. The Open Source brand also runs Packt's Open Source Royalty Scheme, by which Packt gives a royalty to each open source project about whose software a book is sold.

Writing for Packt

We welcome all inquiries from people who are interested in authoring. Book proposals should be sent to author@packtpub.com. If your book idea is still at an early stage and you would like to discuss it first before writing a formal book proposal, then please contact us; one of our commissioning editors will get in touch with you.

We're not just looking for published authors; if you have strong technical skills but no writing experience, our experienced editors can help you develop a writing career, or simply get some additional reward for your expertise.

Testing and Securing Android Studio Applications

ISBN: 978-1-78398-880-8 Paperback: 162 pages

Debug and secure your Android applications with Android Studio

1. Explore the foundations of security and learn how to apply these measures to create secure applications using Android Studio.

2. Create effective test cases, unit tests, and functional tests to ensure your Android applications function correctly.

3. Optimize the performance of your app by debugging and using high-quality code.

Android Studio Application Development

ISBN: 978-1-78328-527-3 Paperback: 110 pages

Create visually appealing applications using the new IntelliJ IDE Android Studio

1. Familiarize yourself with Android Studio IDE.

2. Enhance the user interface for your app using the graphical editor feature.

3. Explore the various features involved in developing an android app and implement them.

Please check www.PacktPub.com for information on our titles

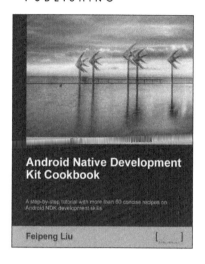

Android Native Development Kit Cookbook

ISBN: 978-1-84969-150-5   Paperback: 346 pages

A step-by-step tutorial with more than 60 concise recipes on Android NDK development skills

1. Build, debug, and profile Android NDK apps.
2. Implement part of Android apps in native C/C++ code.
3. Optimize code performance in assembly with Android NDK.

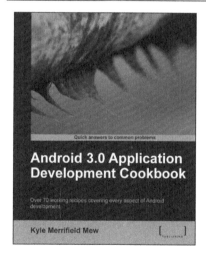

Android 3.0 Application Development Cookbook

ISBN: 978-1-84951-294-7   Paperback: 272 pages

Over 70 working recipes covering every aspect of Android development

1. Written for Android 3.0 but also applicable to lower versions.
2. Quickly develop applications that take advantage of the latest mobile technologies, including web apps, sensors, and touch screens.
3. Discover tips and tricks for varied and imaginative uses of the latest Android features.

Please check **www.PacktPub.com** for information on our titles

CPSIA information can be obtained at www.ICGtesting.com
Printed in the USA
LVOW01s0722300815

452071LV00016B/249/P